CALL DIMENSIONS

Options and Issues in Computer-Assisted
Language Learning

ESL & Applied Linguistics Professional Series
Eli Hinkel, Series Editor

CALL DIMENSIONS

Options and Issues in Computer-Assisted Language Learning

Mike Levy
Griffith University, Australia

Glenn Stockwell
Waseda University, Japan

Routledge
Taylor & Francis Group
New York   London

Routledge is an imprint of the
Taylor & Francis Group, an informa business

First Published by Lawrence Erlbaum Associates, Inc., Publishers
10 Industrial Avenue
Mahwah, New Jersey 07430

Reprinted 2008 by Routledge

Routledge
Taylor and Francis Group
270 Madison Avenue
New York, NY 10016

Routledge
Taylor and Francis Group
2 Park Square
Milton Park, Abingdon
Oxon OX14 4RN

Cover design by Tomai Maridou

Library of Congress Cataloging-in-Publication Data

Levy, Mike, 1953–
    CALL dimensions : options and issues in computer assisted language learning / by Mike Levy and Glenn Stockwell.
        p. cm. — (ESL and applied linguistics professional series)
    Includes bibliographical references and index.
ISBN 0-8058-5633-1 (cloth : alk. paper)
ISBN 0-8058-5634X (pbk. : alk. paper)
1. Language and languages—Computer-assisted instruction.
    I. Stockwell, Glenn.   II. Title.   III. Series.
P53.28.L483   2006
418.00285—dc22   2006003091
                                                          CIP

Printed in the United States of America
10   9   8   7   6   5   4

Contents

Preface

The field of computer-assisted language learning (CALL) has developed and evolved rapidly over recent years. It now includes an increasingly diverse range of work relating to the principled application of new technologies in language learning. Books are regularly published on CALL, and at least four international journals are now dedicated to the topic. CALL is also represented through organizations and annual conferences around the world, such as CALL and EuroCALL in Europe, CALICO and IALL in the United States, JALTCALL in Japan, and the WorldCALL Conference that, so far, has been held in Australia (1998) and Canada (2003).

Not surprisingly, this volume of activity has led to the production of a sizable corpus of work. In fact, the breadth and diversity of CALL is frequently underestimated (see Egbert, 2005). One of the real problems for the language teacher, software designer, or researcher who wishes to use technology in second- or foreign-language education is how to absorb and relate what has been achieved so far, and how to make sense of it. The kind of understanding that comes from a critical reading of a substantial literature in order to develop a balanced and detailed knowledge of the field is not easily achieved.

CALL Dimensions has been designed to address this problem. The book looks in depth at seven important dimensions of CALL: design, evaluation, computer-mediated communication, theory, research, practice, and technology (chaps. 2–8). Each of these chapters is divided into two major sections: description and discussion. The description section reviews the recent literature, identifies themes, and selects representative projects to illustrate the dimension in question. The discussion section provides in-depth analysis. These two sections are followed by a conclusion, which offers suggestions for further work. The book gives detailed references and links that connect the description and discussion with original works and primary sources, so the reader is able to follow up easily on areas of personal interest. In this book, the label *CALL* is interpreted broadly and is taken to include technology-enhanced language learning, network-based language learning, Web-enhanced language learning, and information and communication technologies for language learning (see Levy & Hubbard, 2005). Also, unlike many other publications in the field, this work deals with a range of languages, rather than just English.

The target audience for the book includes the following five groups: the independent researcher, developer, or practitioner reading for a broader and deeper understanding of CALL; language teacher-designers, who are increasingly required to create online materials for independent study or distance learning; those learning about the use of technology and language learning in the increasing number of CALL teacher education programs being conducted around the world; CALL researchers, ranging from those who conduct classroom-based action research to those who engage in large-scale research projects; and those involved in evaluating technology and language learning, from classroom-based evaluation studies to online courses, Web sites, and tasks.

THE ORGANIZATION OF THE BOOK

There are 10 chapters in this book: an introduction, seven chapters that cover the dimensions of CALL, and two concluding chapters that complete the book. The first chapter sets the scene and introduces major areas of interest and growth in CALL. These areas include new technologies and their purpose, evolving ideas of what language learning is thought to involve—especially in the form of tasks—and the impact on CALL of factors related to the learner, the learning environment, and the target language itself.

Chapter 2 focuses on design as it relates to Web sites, online courses, tasks, and collaborative projects for a number of languages. It looks especially at points of departure and the stages in the design process. This leads to a discussion of ways in which CALL design can be improved. Evaluation goes hand in hand with design, and thus it naturally follows as the focus in chapter 3. This chapter discusses, with examples, the wide range of new technologies and new pedagogies being evaluated in CALL today. In looking at these various aspects of evaluation, the chapter reflects on criteria and principles for evaluation, and analyzes some well-known frameworks. One area that requires special attention is computer-mediated communication (CMC) and CALL, and this provides the theme for chapter 4. This chapter reviews the current use of CMC technologies for language learning, and examines their individual strengths and limitations.

Chapter 5 looks in detail at the theories that have been employed in CALL. It briefly describes their philosophical bases with key concepts, and gives extensive references. The chapter concentrates primarily on the interaction account of SLA, sociocultural theory, activity theory, and constructivism. These different theoretical perspectives are assessed in terms of their focus, scope, and range of application, as well as their strengths and limitations for CALL. The chapter also looks at the role of theory and the designer, researcher, and teacher as consumers. Leading on from the theory dimension, chapter 6 examines research, and is organized around six clearly identifiable research strands in CALL. These have been carefully chosen to represent important research directions currently being pursued in the field. Each research strand is represented by one specific research study that is described in detail. Collectively, these studies cover a range of research topics, designs, and methods used in CALL.

Every day, CALL practitioners are dealing with the realities of technology adoption and use. Language teachers employing technology in some way are faced regularly with the practicalities of making it a motivating and effective learning experience for students. Chapter 7 concentrates on the actual practice of CALL. The chapter includes a description of the types of activities that are used to develop the language skills—including listening, speaking, reading, and writing—as well as language areas, including grammar, vocabulary, and pronunciation. Chapter 8 focuses on current advances in technology and their effects on the way that we think about, select, and use technology in language learning. Each technology is described in terms of its relevance to language learning, and its strengths and shortcomings.

The book closes with two concluding chapters that discuss how the various dimensions might be brought together, the first from a practical point of view, the second with a view to the development of CALL as a whole. Chapter 9 examines the important question of integration—what integration really means and how it might be accomplished in institutional settings such as schools and universities. The chapter also considers the role of the teacher as designer. Finally, chapter 10 marks a distinction between emergent CALL and established CALL. Emergent CALL is identified as an area in which practitioners concentrate on testing state-of-the-art technologies and developing innovative programs and practices. Established CALL utilizes tried and tested mainstream technologies in which the emphasis is on developing good pedagogy and practice, and designing effective language-learning tasks and projects. Thus, chapter by chapter, the book creates, through description, analysis, examples, and discussion—a detailed picture of modern CALL.

—*Mike Levy*
—*Glenn Stockwell*

Acknowledgments

Our collaboration on this book began some years ago, although the idea really began to gather momentum at the 2003 EuroCALL Conference in Limerick, Ireland. Countless discussions followed, mostly thanks to the technologies of phone and e-mail, as we continued with our work in Australia and Japan. This collaboration has been a rewarding experience for both of us.

We would like to express our gratitude to our colleagues in CALL who have contributed in so many ways in the writing of this book. We would especially like to express our appreciation to Philip Hubbard (Stanford University) and Martha Pennington (Elizabethtown College) for their very perceptive and detailed comments on drafts of the manuscript. Thanks are also due to the Centre for Applied Language, Literacy and Communication Studies (Griffith University) for their assistance. We would like to thank the series editor, Eli Hinkel, for her guidance, and Naomi Silverman, Erica Kica, and Marianna Vertullo at Lawrence Erlbaum Associates, as well as copyediter Gale Miller, for their helpful support in allowing the book to reach completion. Finally, we would especially like to thank our families for their patience and support throughout the project.

List of Figures and Tables

FIGURES

TABLES

Chapter 1

Introduction

During the last 3 decades, computer-assisted language learning (CALL) has progressed and evolved at a remarkable rate.[1] Books are regularly published on the subject and at least four international journals are now dedicated to the topic. There are annual conferences devoted to it in many parts of the world, and highly active online discussion lists. This activity amply demonstrates a strong interest and commitment to the field and the range of work being undertaken. A closer examination reveals a steadily increasing level of specialization, especially in efforts to refine the underlying philosophy and principles that support CALL design, evaluation, research, and practice. CALL has grown along both the horizontal and vertical axes: It has become a rich and diverse area of work with considerable depth.

The richness and diversity of contemporary CALL, when viewed in its entirety, is a result of many factors. These include the number and range of technological tools available with the potential for use in CALL applications; an increasingly sophisticated understanding of how languages are learned (although lacking a single, overarching theory to rely on as a guide); environmental factors that lead to a variety of priorities, resources, and objectives for different learners in different settings; and particular challenges that arise as a result of the attributes or qualities of the target lan-

---

[1]Although many other acronyms have been suggested, CALL is the one that seems to have stuck. It has become bigger than any particular meaning that might be ascribed to it, and has been longer-lasting and more widespread than any comparable acronym or term—for example, NBLT (network-based language teaching); TELL (technology-enhanced language learning), or information and communication technologies for language learning. CALL is also used in regular annual conferences around the world and in the leading journal titles. As a result, it is the acronym/term used in this volume (for further discussion, see Levy & Hubbard, 2005).

guage. There are other factors as well, but by way of introduction it is worth discussing these four areas first of all.

In the developed world, the array of communication technologies now in use provides a good example of how technology has diversified and evolved in recent years.[2] Aside from face-to-face communication, we might choose to communicate with our family, friends, or colleagues via phone (mobile or landline; voice, text, or images), or via e-mail or chat, either one to one or as part of a group (such as a tele- or videoconference). These developments in the wider world are reflected in our educational institutions, albeit with some delay. Clearly, language teachers and learners have an increasing number of possible options, whereas the use of any particular tool for teaching or learning requires a clear sense of its strengths and limitations, and an understanding of how to match the qualities of the tool with suitable language-learning tasks.

Historically, the invention of new technologies has been largely motivated by a desire to extend or to overcome our innate limitations as human beings, especially those set by our physical or mental capabilities. Thus, new technologies such as the pen and paper for writing aid our memory, the telescope or microscope enhance our vision, the telephone extends our ability to communicate at a distance, the hammer amplifies our strength, and the car or airplane extend our range. Interestingly, many of the most penetrating technologies introduced in the last 30 years have extended our ability to communicate with people at a distance, or at times when they are not immediately available. Language is intrinsically a part of these developments, because the attributes (or affordances) of each technology help shape how interactions take place and how language is used in each setting. This is just as true for nonnative speakers of a language as it is for native speakers.

The technologies used in CALL extend well beyond communication tools, however. They also include generic tools and devices such as the word processor for writing, online dictionaries for vocabulary work, or MP3 players for intensive listening practice. Typically, when compared to the more conventional alternatives, these tools are useful in providing a means for manipulating language more effectively (e.g., the word processor), supplying context-sensitive help or information more promptly (e.g., pop-up word definitions or examples), or otherwise supporting the processes required for language learning, frequently by offering greater flexibility (e.g., language practice at a time and place that suits the learner). In each case, the technological tool should be considered in relation to the task for which it is intended. Ideally, a suitable pedagogy is de-

---

[2]The terms *technology* and *technological tools* are used in this book to refer to both hardware and software.

vised to ensure that the CALL materials are used in an appropriate, principled, and effective way.

The application of these technologies is often direct, but that is not always the case. Technological tools are also used indirectly, as a means to an end; for example, in the construction and testing of new CALL materials. This activity is one of the unique, defining features of the field. In this book, the term *CALL materials* is used to include the wide range of CALL artifacts or products that language teachers and designers create using technological resources (see Levy, 1997). The term *CALL materials* is used to encompass tasks, software, courseware, Web sites, online courses, programs, packages, and learning environments. This label is used to emphasize the connection between CALL and language-learning materials development in general—where the term *materials* is the accepted term (see Tomlinson, 1998). A sense of continuity between CALL and language teaching more generally is advantageous, especially in relation to language-learning materials design and development. Although in some instances materials and learning environments are distinguished and treated separately, learning environments on the computer are generally included under the materials umbrella too.[3]

From its earliest days, those in the field have been writing CALL materials for themselves, or working closely with others. CALL is about design, development, and evaluation, as well as research and practice conceived around a ready-made product or generic application, such as an e-mail or word processing program. In designing and developing new materials, the teacher or designer has alternatives. Prospective authors may turn to the use of a general-purpose Web editor (e.g., *FrontPage*), or a mainstream multimedia application like a word processor or a presentation tool (e.g., *Word, PowerPoint*) for development work. Alternatively, they may choose to work with a learning management system (LMS), which, in essence, is a Web-based software application that provides the teacher with an integrated system for distributing course materials, communicating with students, instigating student–student discussions, presenting quizzes, and managing a range of administrative tasks. Good examples of an LMS are *BlackBoard* and *WebCT*. Such systems are becoming increasingly common develop-

---

[3]This follows the early work of Breen, Candlin, and Waters (1979, p. 5) who, in the case of communicative language teaching (CLT), suggested the development of two kinds of materials: *content* materials as sources of data and information; and *process* materials to serve as "guidelines or frameworks for the learners' use of communicative knowledge and abilities" (Breen et al., 1979, p. 5). First, learning environments on the computer were likened to process materials in that they provide frameworks within which learners can use and practice their communicative skills. The notion of materials as guidelines or frameworks for learning was reinforced by Allwright (1981), who argued for materials to be related to the "cooperative management of language learning" (p. 5). Learning environments on the computer fit comfortably within this broad definition of materials.

ment tools at universities around the world. Another option is to use an authoring tool, such as *Hot Potatoes*, which provides easy-to-use templates for developing different kinds of CALL activities. Those who are especially dedicated and with the requisite expertise may turn to a markup language, which contains a wide variety of codes that can be attached to a text to indicate layout, styling, and interpretation when the document is viewed by a special application. Whichever path is chosen, the options are many, and the author's or designer's choice of an authoring tool needs to be suitably informed and guided.

Often, the result of such efforts is a language-learning Web site. Web sites may aim to serve a wide audience (e.g., all skills from beginner to advanced), or a more narrowly defined one (e.g., listening practice for beginners). CALL sites can differ greatly in scale and focus. They can be straightforward and simple in their aims, or multifaceted and surprisingly diverse in their range and functionality, sometimes including a large number of activities and functions supported by a range of technologies. An exceptional example of what can be achieved is the *Learn Welsh* Web site (BBC: Learn Welsh, http://www.bbc.co.uk/wales/learnwelsh/). The site is multidimensional and includes informal and more formal opportunities for learning language at levels from entry to advanced. It incorporates the social and cultural dimension of language learning with many examples, and includes links to many other Web sites focusing on the Welsh language or culture. It includes activities and resources ranging from the quick and simple, such as the catchphrase of the day, to the more sophisticated. For example, in the 3-D virtual town, the user can interact in language or video scenarios aimed at Welsh in the workplace and undertake accompanying exercises. Online dictionaries, spell checkers, and grammar exercises are also provided. The site makes full use of new technologies, including computer-mediated communication (CMC), such as the message board, and the text club that sends themed phases three times a week directly to the user's mobile phone. This site illustrates what can be accomplished with imagination, a high level of resourcing, and the political will to maintain and support a minority language.

Alongside these technological developments, there have been increases in our understanding of language and how it is learned. We now have a much expanded and refined body of research to support our work. However, although we now know more about language acquisition, this new knowledge has not led to a single unified theory of language learning, but instead to a number of more narrowly defined theories that focus on specific questions and aspects of language and language learning (Jordan, 2004; Mitchell & Myles, 2004). Theoretical perspectives have also increasingly become more attuned to individual and social factors that govern the successful acquisition of a foreign or second language. For the CALL practitioner—who might be considered a consumer looking toward theory for guidance—the circumstances are not as straightforward as they once were. There are a number of compelling theories to choose from, drawn from second language acquisition research, education, psychology, and human–computer interface design. The theories have grown not only in number, but also in sophistication and complexity. Some, such as sociocultural theory, involve a considerable number of specialized concepts and levels of analysis. Thus, although using theory as a point of departure is generally to be recommended, there is no doubt that in opting to proceed in this way—in a principled fashion led by theoretical insights—the scale and complexity of the challenge has been increased. What is now needed is not so much a single solution to guide CALL, but rather a careful weighing of the options so that strengths and limitations become evident. This viewpoint grows from a belief that postgraduates, researchers, and teachers increasingly have to relate to a number of parallel perspectives in their work. This requires an appreciation and understanding not of one single viewpoint alone, but instead of a number of different perspectives concerning the nature of language and language learning.

In practical terms, our developing understanding of language teaching and learning is well represented in evolving conceptions of the language-learning task (Long & Crookes, 1991; Nunan, 2004; Skehan, 1998; Willis, 1996). Definitions of tasks have changed significantly over the last 20 years (e.g., Candlin & Murphy, 1986; Johnson, 2003; Ribé & Vidal, 1993). According to Ribé and Vidal (1993), task goals have developed from the narrower objective of activating communication and cognitive strategies to a much broader view of enriching the students' whole experience of language and language learning. Thus, tasks may now seek to develop the learners' motivation, their creativity, or their awareness of the many aspects that make up a language, such as the cultural or pragmatic dimensions. We might also add to this list the goals of developing learning strategies and learner autonomy. Clearly, language-learning tasks are not what they were. These new conceptions of language-learning tasks lead language teachers and designers to be responsive to a number of objectives in task design. Pedagogical goals may be partially achieved through a single task but, more likely, this combination of goals and objectives will lead to the design of a sequence, or a cycle of tasks that aims to address different goals at different times (see Levy & Kennedy, 2004; Willis, 1996). As a result, greater pedagogical expertise is now required in structuring the language-learning experience for students and in designing tasks that are appropriate in meeting learner needs and aspirations. The work is challenging for face-to-face teaching in the language classroom. It is even more so in CALL environments, where a larger number of factors are involved because of the range of technological options available and the need to effectively manage and integrate CALL work with non-CALL work.

The choice and use of technological tool/s and our understanding of how a language is learned play a large part in governing any conceptualization of CALL. Further critical factors derive from the nature of the learner and the learning context. These factors include consideration of specific learner characteristics (background, needs, goals), features of the setting (classroom, lab, home), and factors concerning the learning environment (technological, societal, cultural, institutional). The target language, the curriculum, and the teacher(s) have to be taken into account as well. These various context-specific factors help shape and determine any interpretation of CALL, and they need to be understood (see Levy, 1997). As before, we see a complex mix of variables that need to be negotiated in order to arrive at an effective CALL implementation.

To give just one example, consider for a moment a young native speaker of Japanese learning English in the latter years of high school in Japan. If we were to contemplate how CALL might be implemented in this situation, we would need to consider, at the very least, the following:

- Japan's physical distance from English-speaking countries.
- The traditional approach to education and learning in Japan.
- Cultural goals and expectations.
- Large, homogeneous classes.
- The critical importance of success in English, especially for high school students in relation to the university entrance examinations.
- The role that university examinations play (they largely govern high school curricula, especially in later years).
- A generally sophisticated and robust technological infrastructure.
- The ubiquity of mobile technologies, especially mobile phones.

The specific factors and constraints in this profile would contribute to a particular interpretation of CALL in this setting. The point to be made here is that there are many variables to take into account in any given setting, and the decision-making process is multifaceted.

Finally, the nature of the target language itself should not be overlooked. It is surprising how often the language in question is assumed to be English. Many papers and books on CALL, knowingly or unknowingly, restrict themselves to this particular language perspective. Of course, the learning of English as a second, third, or foreign language is very important because of the significance of English as a world language. For many language learners around the world, English is the target. Also, there are many native or near-native speakers of English. This means that there are often native speakers available for collaborative projects such as those undertaken in the various kinds of CMC-based CALL. In practice, however, CALL spans many more languages than English. Often, technologies are harnessed in rather different ways, depending on the qualities and characteristics of the target language and the background and goals of the students who wish to learn them. Good examples are the scripted languages such as Japanese and the tonal languages such as Mandarin Chinese, which have specific technological requirements such as two-bit character fonts and tonal diacritics. This book is not restricted to the learning of English alone.[4] We look at CALL applications and include examples in the learning of many different languages as well as English (although English still remains a very important focus).

It is clear from this short, introductory discussion that CALL consists of a rather large body of work with a sizable amount of information to absorb, and pedagogical alternatives that require matching the tool to the task and choosing from a wide range of options in relation to design, theory, and practice. The range of technological tools, and the many ways they may be applied in language learning plus the challenge of designing appropriate CALL tasks to meet the needs of learners in different settings, combine to create a considerable degree of complexity because decisions are required across a number of areas and levels.

APPROACH AND RATIONALE

This book has been designed to address the diversity and complexity of modern CALL. It seeks to describe the major topics and developments in CALL, with examples to illustrate key ideas, themes, and directions. Importantly, it seeks to embrace a degree of complexity and alternative viewpoints, rather than avoiding differences of opinion and attempting to recommend, force, or imagine a single perspective for the work of the field. We do not feel that there are grounds to support a single perspective for CALL, as we try and illustrate through the wide variety of projects described. In order to understand CALL and develop new CALL materials, we believe the best way to proceed is to consider a sample of well-conceived projects with the practical and theoretical arguments that motivate and justify them. This approach takes into account the complexities of language learning, the absence of a grand theory, and the modular approach gener-

---

[4]An interesting example of the dominance of English on the Internet is in Web site domain names (i.e., the actual Web site address). Until recently, domain names had to be composed using the 26-letter Latin alphabet plus hyphens, underscores, and the digits 0 to 9. In the United Kingdom, a not-for-profit company called Nominet has challenged this dominance with plans to introduce international domain names (IDN) that permit the inclusion of accents or entirely different alphabets, such as Arabic (see Wray, 2005).

ally taken toward theory construction and research in second language acquisition (Mitchell & Myles, 2004).[5] The approach is also very relevant for CALL because it takes into account the wide variety of technological tools available, each with their strengths and limitations, and the many goals and priorities held by language teachers and learners today. Overall, we believe that dealing with complexity and multiple viewpoints is a necessity in CALL at this time. This book does not claim to be able to provide simple answers, but it does offer a detailed background on the alternatives, and a foothold for the reader to use as a basis for informed decision making.

The approach taken in the book is essentially inductive, or data driven. Chapters 2 to 8 begin with key points concerning the CALL dimension in focus. They continue by describing significant developments in the area with a few carefully chosen and observed exemplars from the literature (the description section included in each of these chapters). Then, on this basis, the chapters draw out themes, discuss prevailing approaches and issues, and, when called for, introduce new ideas and concepts (the discussion section). Sometimes the descriptions are quite detailed, because any CALL implementation is heavily influenced by contextual factors. More details help provide a better understanding of the context, which, in turn, more readily allows readers to draw parallels between the example and their own situations. The aim is to build on previous work, absorb the ideas and directions being taken, and produce new understandings on this basis, rather than to convince readers of the superiority of one particular approach or viewpoint. We do not try to gloss over contradictions and ambiguities that may exist in the field, or the different ways in which designers, teachers, and researchers approach the question of how best to make use of new technologies in language learning. We prefer to see what emerges as significant in terms of how the people in the situation construe it. In other words, we look at CALL from within, as a worthy body of work in its own right, and represent it as best we can. Sometimes this leads to perspectives and approaches that are unique to CALL.

Initially, the book was motivated by a research project whose aim was to describe the breadth and depth of CALL in a way that was systematic rather than anecdotal. Thus, a large and representative sample of recent publications was collected to form a CALL corpus. The corpus was described in a systematic way, using a specially designed set of keyword descriptors (a CALL thesaurus). Through an inductive process of annotating a large number of CALL publications, patterns of CALL work were identified. Goals, di-

---

[5]The label *theory* is applied broadly in this book to signify a set of ideas or general principles that describe or explain an aspect of language or language learning. The term is held to encompass an account, model, framework, or hypothesis. Examples include the interaction account, constructivism, flow theory, situated learning, and task-based language teaching, described by Doughty and Long (2003, p. 51) as an "embryonic" theory of language teaching.

rections, methods, and procedures were also clarified. Levy (2000, 2002) described this study in detail in relation to CALL research (2000) and design (2002). (This study is also discussed in appendix A.) Although the CALL corpus provided the initial stimulus for the book, this was only a beginning. A wide range of recent book and journal publications in CALL and related areas have been added to the dataset, and they also provide the basic and essential reference materials for the development of the ideas presented here.

The overarching structure of this book derives from the empirical study. The chapter titles and the basic ordering of the chapters correspond to keyword descriptors or "identifiers" that were used to describe the literature (see appendix A). We have called these major identifiers "CALL dimensions" and, in order, they are: design, evaluation, CMC, theory, research, practice, and technology. As such, there is an empirical basis for the chapter headings and their sequencing, and grounds to say that the content really does reflect the scope and interests of CALL as an emerging, semi-autonomous discipline. Particularly noteworthy in this respect is the importance of design, so this is where we begin in the next chapter.

Design

In many respects, language teachers may be considered designers. Not only do many language teachers design or adapt materials, and develop tasks and courses to match the needs and goals of their students (online and offline), they are also designers in the way they organize and manage their classes, programs, time, and resources. The role of the language teacher as designer is often underestimated or overlooked; this is noted and discussed at points throughout the book (a more in-depth discussion of the role of language teacher as designer presented in chap. 9). Design, in one or another of its aspects, is also a recurring theme in CALL. Design permeates the field and has been a central motif in CALL publications over the last 20 years. It has arisen repeatedly in terms of materials design, task design, syllabus design, course design, exercise design, instructional design, and screen design.[1] Design enters into the discourse of CALL in many forms and at a variety of levels, from the scale of an institution down to the level of an exercise. For example, Barr and Gillespie (2003) reflected on design at the level of the institution and compared the pedagogical effectiveness of the computer-based learning environments at three universities in relation

---

[1]Like design, the idea of *development* has also been a central theme in CALL since its earliest days. Its presence is especially visible in relation to materials development (both CALL and non-CALL), and in the activities of professional organizations such as CALICO's Courseware Development Interest Section and TESOL's Developer's Showcase. Although there are many variations, the process of creating new learning materials for delivery by computer has traditionally been divided into design, development, and implementation, where the design stage refers to the initial conceptual planning phase, development refers to the building and refinement that occurs next, and implementation involves issues concerned with the actual use of the new materials in practice (see Dabbagh & Burton, 1999; Levy, 1997; Shneiderman, 1987). For the purposes of this book, these three elements are not strictly separated. Development issues, when they arise, are largely incorporated into the design chapter, although they appear in other chapters too, as in the evaluation chapter, for example, where it is common to see the

to CALL. In contrast, working within narrower boundaries, Sivert and Egbert (1999) described the design of a computer-enhanced language classroom, and Strambi and Bouvet (2003) described the thinking behind two online distance language courses in Italian and French for beginners. Other examples include the design of Web sites, tasks, and activities, moving to perhaps the smallest unit of consideration, the exercise, as exemplified by Shawback and Terhune (2002), who used an authoring tool to design a series of online interactive exercises. This range of products and objectives reflects the many goals and orientations of CALL designers.

This diversity in level and focus is reinforced when the points of departure for projects are considered. A small sample from the literature is indicative of the range and points of focus (italics added):

- Given these tendencies, my concept for the *design* of foreign language instructional software derives from the need to achieve an optimal mix between in-class and out-of-class learning. (Pusack, 1999, p. 26)
- In this paper, we have explored the *design* evolution of both software and pedagogy in the creation of IRC Français. (Hudson & Bruckman, 2002, p. 130)
- [W]e *designed* and developed learning materials and tasks to be distributed on CD-ROM, complemented by a *WebCT* component for added interactivity and task authenticity. (Strambi & Bouvet, 2003, p. 1)
- In the preceding discussion we have tried to show how the *design* features … are motivated by theoretically driven empirical research. (Van de Poel & Swanepoel, 2003, p. 206)
- The first part of this article presents the steps necessary for *designing* an effective language learning tool to foster communication and negotiation, taking into consideration the importance of supporting integral education, using tasks, providing elaborated input and feedback, and promoting collaborative learning. (González-Lloret, 2003, p. 86)
- The project *design* involved the creation of a customised online course management shell incorporating HTML, Macromedia, Flash and Cold Fusion technologies. (Ayres, 2003, p. 351)

This short selection of excerpts immediately illustrates some of the prevailing topics that arise when CALL practitioners talk about design. They demonstrate the importance of integrating CALL work with non-CALL work, for instance. As Pusack (1999) observed, at the class level, in-class and out-of-class work need to be successfully combined or integrated. As differ-

evaluation results feeding back into subsequent development. It is still assumed that design generally precedes development, although in actual practice design and development are not so readily separated; this is because technical obstacles and trade-offs during the development process typically require further design decisions to be made. Note that in the original CALL corpus, the keywords *design* had 93/177 hits and *development* had 34/177 hits (see appendix A). Finally, note that *implementation* is also not considered as an item in its own right; in this book, issues concerning implementation are covered in chapter 7.

ent technological resources are often available in class and out of class—in the library or home, for example—the integration of these elements needs to be thoughtfully and coherently designed, often with the needs and resources of the individual learner in mind.

The points of departure for a design vary significantly. Sometimes a design will be theory driven, sometimes the nature of a particular project or task will be the primary idea or concept that shapes the design, and at other times the parameters of the development environment will take precedence. For example, if language-learning materials are to be distributed via the Web, or on a CD, or developed using a learning management system (LMS) such as *WebCT* or *BlackBoard*, then these decisions inevitably shape the design in important ways, irrespective of other aspects such as the theoretical or pedagogical motivation. It is also worth noting that in the Strambi and Bouvet (2003) and Ayres (2003) excerpts, typically in contemporary CALL designers are not using one single technology but rather a number of complementary technologies in their work.

The point where one begins in design is critical. Whether it is a theory, pedagogical model, course or syllabus, task, exercise, language skill, technology, or some kind of mix, the whole design unfolds from that point on. In other words, the point of departure sets the compass and the direction one will take. The final product will be largely determined by the initial decisions that are made. An assessment of student needs is also a frequent consideration at the initial stages in a new CALL project.

However, it is not only a question of what is foremost in the designer's mind at the outset; it also a question of what design process is adopted and what decisions are made as the process unfolds. Typically, trade-offs have to be made because of conflicts between what the designers might like to do and what they are actually able to do. The statement by Hudson and Bruckman (2002) showed that design is not a static entity that is set once and for all at the outset of a new project. Designs in CALL evolve and change as designers come to understand the detail of the pedagogical motive and the particular opportunities and constraints provided by the technological resources that are available. Also, depending on the particular design process that is adopted (see Hémard, 2003), there may be opportunities to collect significant amounts of information about the users or language learners, such information can be fed back into the design to clarify and improve its focus and direction. Each excerpt in the preceding list illustrated the centrality of design in the initial conceptualization of a CALL project or task, and the very important point that CALL design is heavily dependent on context. In other words, designers are often concerned primarily with meeting local needs, typically related to their own institution, learners, or curriculum. The path eventually taken is dictated by many factors, not least of which involve the skills, imagination, and resourcefulness of the designers, both technically and pedagogically.

From this brief and preliminary snapshot of CALL design, it is already clear that design is a multifaceted phenomenon and that language teacher-designers talk about their work at a number of levels and with a variety of focal points. Consequently, there are many ways in which one might approach a discussion of this topic.

This chapter begins with a selected sample of the products of CALL designs. A range of CALL materials is included to give the reader a sense of the scope of activity in CALL design and the many points of departure. These primary examples have been selected for three principal reasons:

1. They are each a strong and convincing example of their type. Thus, if the design of an online course or an online tutor is described, the example selected is a good one. Its design has been well conceived and implemented, and the project has been developed using clear and justifiable principles.
2. They each help to form, as a group, a representative cross-section of approaches to CALL design and design issues, in terms of:
   • Point(s) of departure.
   • Theoretical motivation (single theory and mixed).
   • Methodological or pedagogical motivation (a variety given).
   • Course or syllabus orientation.
   • Task orientation.
   • Skill orientation.
   • Technology choice(s).
   • CALL tutors.
   • CALL tools.
   • Integration.
3. They are written by authors who, in many cases, have been working along similar lines over a number of years and have a track record in the area.

In the descriptive section that follows, the broad parameters of design in CALL are sketched out. General ideas and principles in language teaching and learning are introduced, before more CALL-specific ones, to give a better sense of the broader context in which CALL design is being accomplished. Language-learning tasks, course and syllabus design, and pedagogical aspects feature strongly. Then, some of the conceptual frameworks and themes more directly derived from previous work in CALL are covered. At regular intervals throughout this analysis, specific CALL project examples are included. In the discussion section toward the end of the chapter, central themes and issues emerging from the descriptive section are examined, and the implications discussed at greater length.

DESCRIPTION

Language-Learning Tasks

In language teaching in the mid- to late 1980s, publications such as *Language Learning Tasks* (Candlin & Murphy, 1986) and *Designing Tasks for the Communicative Classroom* (Nunan, 1989; see also Nunan, 2004) helped to set a task-oriented agenda and establish the language-learning task as a pivotal component in design (see also Long & Crookes, 1991). Nunan's (1989) book dealt with the design and development of communicative language tasks and saw the task as "a basic planning tool" (p. 1) and "the major focus of the teacher's planning efforts" (p. 134). It is noteworthy that the design of language-learning tasks has remained firmly on the agenda to the present day (see Ellis, 2003; Skehan 1998). More recent examples in CALL include Salaberry (2000b) on the pedagogical design of CMC tasks, Skehan's (2003) article on focus on form, tasks, and technology, and Chapelle's (2003) book, which pays considerable attention to determining principled designs for language-learning tasks.

Nunan provided an early summary of the many definitions of a language-learning "task" (Nunan, 1989; see also Breen, 1986; Candlin, 1986). Nunan defined tasks in terms of "their goals, the input data, linguistic or otherwise ... the activities derived from the input, and the roles and settings for different tasks for teachers and learners" (1989, p. 2). He also added that however well designed and conceived the task, "we cannot know for certain how different learners are likely to carry out a task. We tend to assume that the way we look at a task will be the way learners look at it" (Nunan, 1989, p. 20; see also Breen, 1986). One way of dealing with this tendency is to involve learners in designing or selecting tasks, and this point was taken up at length in Breen's paper "Learner Contributions to Task Design" (Breen, 1986). Among other things, the literature on tasks in the late 1980s illustrated some of the difficulties in reaching an agreed-on decision on a definition of task, the complexities of implementing a task that conforms to a design, and also the value of involving learners in task design.

As far as research on L2 classroom learning is concerned, tasks have also played a very productive role (Ellis, 1994; Larsen-Freeman & Long, 1991). The main research goal related to tasks has been to discover how certain variables affect the interaction that occurs when learners attempt to perform a task (Ellis, 1994, p. 595). The variables can be broadly classified into those that relate to the task and those that relate to the learner. There have been problems determining which variables to investigate and, as a result, a "task framework" has been suggested that can be used to "classify and compare" different tasks. However, as in the earlier examples, definitions of task vary, and the term remains rather vague (Ellis, 1994). In 1994, Ellis maintained, "Attempts to classify task variables in this way are helpful, but we are a long way from developing a taxonomy that is both complete and psychologically justified" (p. 596). Many would agree that this statement is still true today.

The task construct is frequently used as a means of converting a language teaching approach, or a theory of language learning, into a practical activity for students to complete. The task is often held to embody the fundamental principles of the design. Examples from the CALL literature include Chapelle (1999b), Mills (J. Mills, 1999), Salaberry (1996), and Shield, Weininger, and Davies (1999). Meskill (1999) provided a "task anatomy for sociocollaborative language learning" (p. 145) based on Cohen's notion of "multiple ability tasks." These tasks:

- have more than one answer or more than one way to solve the problem
- are intrinsically interesting and rewarding
- allow different students to make different contributions
- use multimedia
- involve sight, sound, and touch
- require a variety of skills and behaviours
- also require reading and writing
- are challenging. (Cohen, 1994, p. 68)

This characterization of task is different again from the ones given earlier. This task profile indicates a strong response to the multimedia elements available in a CALL environment, real-world problem solving, and individual differences.

The prominence of the language-learning task in conceptualizing CALL designs is also evident in much recent work. For example, González-Lloret (2003) described an Internet-based CALL activity for Spanish called *En Busca de Esmeraldas*. The theoretical basis for this design lies in interactionist theory (see Gass, 2003) and a particular interpretation for CALL purposes given by Chapelle (1998; see also Chapelle, 2001). Also, principles underpinning task-based language teaching (TBLT) play an important part in the design of the activity (see also Doughty & Long, 2003; for further discussion, see chap. 5).

Following similar lines, Mishan and Strunz (2003) looked to the nature of the language task to orient and shape the design of an electronic resource book. They described the creation of interactive resources for authentic language learning using *XML*, the *eXtensible Markup Language*, to fulfill this aim (for more on this, see chap. 8).[2] To guide their design, they used a set of principles incorporated into a pedagogical model, referred to as the *authenticity-centered approach*. The notion of authenticity was discussed at some length, both in terms of authentic texts and task authenticity, with some valuable insights. For tasks to be authentic, Mishan and Strunz maintained that they should be designed to:

1. Respond to the original communicative purpose of the text.
2. "Rehearse" real-life tasks.
3. Orient toward the goal/outcome.
4. Create genuine suspense as to their outcome.
5. Require natural (native speaker-like) interaction between learner(s) and the text.
6. Involve genuine communication between learners.
7. Activate learners' existing knowledge of the target language and culture (p. 240).

A task specification is a frequent point of departure and means of framing CALL designs, also employed by Salaberry (2000b) and Jarvis (2001). In many ways, in contemporary language learning the task has come to be the *means* or *agent* of learning. Therefore, the structure, content, and sequencing of language-learning tasks are critical, a point to which we return at intervals throughout the book.

It is clear that *task* is a term that is open to many different interpretations. Definitions are being drawn, not only from the language-teaching/learning domains, but also from education more broadly. Technology elements are clearly influencing these particular realizations. Also, however well theorized or designed the task may be, we do not know how the individual learner will perceive or respond to it. This is a crucial point, made by Hémard (1999), Goodfellow (1999), and others in CALL, and points to research that investigates learner perceptions more fully, and a much closer involvement of the user in the actual processes of task design from the beginning.

At this point in this section on language-learning tasks, it is also appropriate to mention *WebQuests*, which have been widely used in various forms in language learning (Felix, 2002). In an early definition, (Dodge, 1995) noted that a WebQuest is "an inquiry-oriented activity in which most of the information learners work with comes from the Web." According to the online WebQuest taxonomy, "the task is the single most important part" of a WebQuest. It provides a goal and focus for the student and "it makes concrete the curricula in-

---

[2]The Mishan and Strunz (2003) paper is also noteworthy for its clear overview of XML, and some of its special features, especially the way this design tool allows content to be separated from style, thereby providing the design with much greater flexibility.

tentions of the designer" (http://edweb.sdsu.edu/webquest/t askonomy.html). A wide range of tasks is described on this Web site. Although they are not specifically designed for language learning, the tasks could easily be adapted for this purpose. In a paper called "Promoting Student Inquiry: WebQuests to Web Inquiry Projects (WIP)" (Molebash, Dodge, Bell, Mason, & Irving, 2003), the authors described how the WebQuest has been refined as a result of a deeper understanding of the idea of "inquiry." Of WIPs, Molebash et al. wrote:

> WIPs are intended to be used as inquiry roadmaps for teachers desiring to promote higher levels of student-centred inquiry, specifically by leveraging uninterpreted online data to answer inquiry-oriented questions. Unlike WebQuests, which provide students with a procedure and the online resources needed to complete a predefined task, WIPs will place more emphasis in having students determine their own task, define their own procedures, and play a role in finding the needed online resources. (p. 3)

Paralleling recent developments in the definition of the language learning task, WIPs reflect a more determined emphasis on learner autonomy and the learner's role in deciding the questions he or she wishes to answer through using the resources of the Web. Thus, the designer is providing a less structured framework for the students to follow.

Course and Syllabus Design

Although the language-learning task is a frequent point of departure for CALL designers, others begin on a larger scale with their initial conceptual framework, at the level of the course or syllabus. The course may exist already, and the goal may be to add an online component. Alternatively, the goal may be more far-reaching, as when a course is converted in its entirety to an online version, perhaps for distance learners. On other occasions, the course may simply be in the planning stages, and here the CALL component may be planned from the outset. Some examples of CALL designers who have set out in this way include Weinberg (2002), Rogerson-Revell (2003), and Zhang (2002). Weinberg addressed the advantages and difficulties of introducing multimedia technology in an advanced French listening comprehension course. Using work culture and business as broad themes, Rogerson-Revell developed a cultural syllabus centered on European work culture and practices using a mixed "online" and "offline" approach, whereas Zhang provided a comprehensive approach to teaching business Chinese online using courseware that comprises a workbook and a simulation. In each case, the course, or syllabus, is the initial point of focus. In some instances, learners are using the CALL materials alongside a more conventional face-to-face course, whereas at other times learners are working at a distance. Because distance learning courses involving technology are of increasing interest in CALL, we now describe a distance course in a little more depth.

When the language-learning environments are conceptualized and designed to accommodate distance learners, new factors and issues arise and move into the foreground, and new solutions to design problems need to be found. Such was the case with Strambi and Bouvet (2003), who detailed the process of designing and developing two distance courses in Italian and French for beginners. They articulated at some length the differences between on-campus learners and distance language learners, particularly in terms of motivation and teacher–student relationships. They also pointed out the need in this environment to accommodate different patterns of study and ways of engaging students and sustaining learner motivation over time. Theoretically, their point of departure was social-constructivist, although they invoked a certain theoretical pluralism in that they also drew on cognitive and humanistic approaches (see Van de Poel & Swanepoel, 2003, on theoretical and methodological pluralism in design). As we see in chapter 5 on theory, for those designers who present an initial theoretical orientation to their work, some rely solely on a single theory (e.g., Darhower, 2002; Fernández-Garcia & Martínez-Arbelaiz, 2002), whereas others draw on a number of sources that are regarded as complementary, as did Strambi and Bouvet in this example.

Of particular interest in this example is the way in which the authors set about combining the CD and *WebCT* components of the course. This is often referred to as a *hybrid solution* to a design problem, and it usually arises when the designers perceive a specific development tool to be unsatisfactory or in some way limited in a particular area (see Levy, 1999a, 2000; see also Rogerson-Revell, 2003). In this instance, Strambi and Bouvet (2003) decided to make the most of the materials available on CD "to facilitate the distribution of such media rich content, and … to overcome access problems related to bandwidth particularly in remote areas of South Australia" (p. 7). On the other hand, as the authors noted, the CD format imposes limitations in terms of student–student and student–teacher interactivity, and it was primarily for this reason that the *WebCT* component was introduced as well. *WebCT* was used to provide access to a number of communication tools, including e-mail and a bulletin board. It is also noteworthy that the CD interface was developed in HTML and the JavaScript language so that, in the future, the materials might be made available online and distributed over the Internet. In their concluding remarks, Strambi and Bouvet emphasized how much they learned through the design process, which, following Egbert and Thomas (2001), they agreed to be "inherently iterative and evaluative in nature" (p. 404).

Designers are faced with a number of software options to consider when they are deciding which products to use for the development and delivery of their course materials. As we have seen, in designing their course Strambi and Bouvet (2003) chose a CD format and *WebCT* to create the language-learning materials and the environment for their use. Another popular alternative is to use an authoring tool, such as *Hot Potatoes* (Arneil & Holmes, 1999, 2003), or MALTED, another authoring tool of a basically similar kind (see Bangs, 2003). *Hot Potatoes* allows for the rapid development of exercises in six exercise formats for the Web or for stand-alone computer use. A good example was described by Shawback and Terhune (2002), who used the authoring tool to design a series of online interactive exercises to study the language and culture of film.

Methodological Frameworks and Design Integration

It is very clear from the last example that CALL design can be complex, and it requires the careful integration of a number of elements, both pedagogical and technical, in a principled way. Integration of design elements was obviously a critical area for Trinder (2003) who wrote of "juxtaposing approach, content and technology considerations" (p. 79) in the title of her paper. She went on to describe the conceptualization and design of a large-scale multimedia courseware project at Vienna University called the *Online English Mentor* (OEM). Concerning the interrelations among approach, content, and technology-based variables, Trinder maintained that "some of these variables are predetermined by the educational context, whereas others reflect the developers' views of what constitutes an optimal language learning environment" (p. 79). In the OEM, learners can work on reading, vocabulary, grammar tasks, and listening activities. These activities in turn relate to 10 business-oriented core themes that are "meant to build a bridge between general English and the content and terminology oriented ESP of the following semesters" (p. 90). Again, integration of elements is clearly apparent here, this time in terms of content, in relating the OEM work to the wider scheme of work planned for the students.

Trinder's (2003) framework for design is based on the well-known methodological framework introduced by Hubbard (1987, 1988, 1992, 1996). This framework has considerable value in relation to design, development, and evaluation. (It is also discussed at some length in chap. 3 on evaluation.) Hubbard's framework is hierarchical and operates at the three levels of approach, design, and procedure. In short, Hubbard proposed that thinking about design at these three levels can help the designer move from a theoretical orientation or set of principles to a practical, working set of CALL materials (e.g., a Web site, CD, or software program). In this instance, with some modifications, Trinder used Hubbard's framework in the design of the OEM. Trinder asserted that her approach to OEM "is not indebted to one particular method, but is partly underpinned by learning assumptions derived from second language acquisition theory and research, and partly motivated by more down-to-earth considerations like individual teaching experience and departmental teaching philosophy and practices" (p. 85). This mix of motivating forces reflects the eclecticism of the design approach taken by many designers in the modern university context. Trinder also pointed to a number of problems in multimedia design, especially in the area of learner control, and explained how this potentially valuable idea is often compromised by poor user navigation, insufficient or badly located signposts through the program, and poor structuring and presentation of content. She also addressed the "myth" of interactive feedback, and responded to the "multimodality" argument, which says that dual delivery modes (e.g., sound plus vision) are indisputably the most effective for all learners. Trinder's commentary is most useful in these respects.

Language-Learning Areas and Skills

Beyond the task and the course, CALL designers have also been drawn to the language skills and areas when conceptualizing the focus and scope of their work. For this group, the specific orientation is located around a language skill (i.e., speaking, listening, reading, writing), or a language area (i.e., pronunciation, grammar, vocabulary, discourse). This is quite a common occurrence in CALL, reflecting the support roles that the technology can usefully play in these more discrete domains of language learning and the strengths and shortcomings of the technologies involved (see chap. 7 on practice for extended discussion). Recent examples, with their focus, are given in Table 2.1.

In cases such as those highlighted in Table 2.1, the literature and the research associated with the language skill or language area provide the backdrop to the discussion, be it in relation to language teaching with technology, research, or design. Anyone commencing a CALL materials development related to one of these language skills or areas would be well advised to read the corresponding articles before they begin.

In the focus example chosen here to represent the language skills and areas, Van de Poel and Swanepoel (2003) provided an example on vocabulary learning, and presented a design for effective lexical support through CALL. The authors drew on a number of specific theories and methodologies for vocabulary learning. Thus, the theories and methodologies that were employed tend not to be of a general nature (e.g., interactionist, cognitivist, sociocultural). Instead, they were of a specific kind relevant to vocabulary and vocabulary learning. This modular approach to theory is

TABLE 2.1

Language Skills and Language Areas With Recent References in CALL

Skills and Areas	References
Language Skills	
Speaking	Harless, Zier, and Duncan (1999), Nunan (2005), Payne and Whitney (2002)
Listening	Hew and Ohki (2001), Hoven (1999a), Wachowicz and Scott (1999), Weinberg (2002)
Reading	Brandl (2002), Chun (2001), De Ridder (1999, 2000), Dreyer and Nel (2003), Ganderton (1999), Kol and Schcolnik (2000)
Writing	Dodigovic (2002), Glendinning and Howard (2003), Komori and Zimmerman (2001), Pennington (1999b), Schultz (2000), Wible, Kuo, Chien, Liu, and Tsao (2001)
Language Areas	
Pronunciation	Hincks (2003), Kaltenböck (2001), Neri, Cucchiarini, Strik, and Boves (2002), Pennington (1999a), Weinberg and Knoerr (2003)
Grammar	Heift (2001), Salaberry (2000a), Vandeventer (2001), Vanparys and Baten (1999)
Vocabulary	Cobb (1999), Goodfellow, Manning, and Lamy (1999), Greaves and Yang (1999), Jones (1999a), Nesselhauf and Tschichold (2002), Tsou, Wang, and Li (2002), Van de Poel and Swanepoel (2003)
Discourse	Guillot (2002), Kramsch and Anderson (1999), Sotillo (2000)

common for authors who consider their designs from the perspective of a language skill or language area. In this case, the authors commented, "The basic tenet we will adopt is that the design and evaluation of instructional materials should be informed by theoretically driven empirical research on the cognitive processes and representations involved in S/FL vocabulary acquisition" (p. 173). So, not surprisingly, the discussion made reference to those researchers who have focused on understanding the processes of vocabulary acquisition. CALL designers such as Van de Poel and Swanepoel focus on a language skill or language area. As the point of departure, this skill or level tends to be treated as a discrete entity (although not always), with its own literature, noted researchers, and priorities. This is in contrast to mainstream SLA research, which tends to be oriented to the acquisition of grammar, which many SLA researchers regard as the core of the language and most worthy of research and study.

Although the goal of Van de Poel and Swanepoel (2003) was to use empirical research findings regarding vocabulary acquisition as far as was possible, they acknowledged that this was not straightforward. Although rapid advances have certainly been made and partial theories exist—in common with other areas in SLA research—the authors noted that the field is "nowhere near to providing conclusive criteria" for the purposes of design (p. 178). The problem is that to actually use theory in design, one needs very specific, unambiguous information. This idea is discussed further in chapter 5 on theory.

It is also useful to note at this point that the language itself can provide a focus in ways not dissimilar to the language skills and areas. This is especially true when designing CALL materials for languages other than English. The nature of the script might be a key factor, for instance, as in *kanji* learning with Japanese (see Corder & Waller, 2005). For native speakers of English, learning a tonal language such as Mandarin might be the focus in a CALL project because of the difficulties that students face in successfully acquiring the four tones. In this, the relation between the L1 and the target language, the L2, is crucial. Clearly, the situation would be different, for example, if we were designing a CALL package for Korean learners of Japanese. In such cases, knowledge of the learner's L1 and any recognized difficulties for that group in learning the L2 come to our attention. Languages such as Japanese with complex scripts may well lead designers to consider creating programs to address such language-specific problems. Not surprisingly, the right-to-left script of Arabic caused Cushion and Hémard (2003) to pay particular attention to this issue in the design of a CALL package for Arabic.

Design of CALL Tutors

The design and development of CALL tutors has been continuing almost since the field began (see Levy, 1997). With a CALL tutor, a computer program analyzes and evaluates an individual learner's response to a question, and provides feedback on it. Tutors range from the simple to the complex. Simple tutors give simple right/wrong responses to certain question types, such as true/false or multiple-choice, which frequently accompany online reading exercises and listening activities. There are examples of this kind of tutor feedback in many Web sites for language learning (e.g., *Dave's ESL Café*) and in quizzes created by such products as *BlackBoard* or *WebCT*. The Web-based system *WebWiz* also enables this style of question/answer to be created (http://webwiz.hlc.unimelb.edu.au/), but with a little more versatility because it is purpose built for language learning. *WebWiz* is designed to automate the collection and storage of digital media for language practice and testing. The designer–teacher creates listening and reading comprehension exercises around media objects (e.g., audio, video, animation, images, texts) and a multiple-choice quiz is one of the options available in exercise creation.

With the more sophisticated tutors, a more complex algorithm is required that can perform error diagnosis, error correction, and the generation of individualized learner feedback. The CALL software programs or systems that aim to accomplish this task are often referred to as *intelligent language tutoring systems* (ILTS), or *intelligent computer-assisted language learning* (ICALL) programs (see chap. 8 on technology). The label *intelligent*, when applied to a language-tutoring system, really refers to the system's capacity to analyze grammatical input and then to generate error-specific learner feedback (Heift, 2003; Toole & Heift, 2002). In addition, the systems implement natural language processing (NLP) techniques, which utilize theories of grammar to process learner input in order to generate feedback. Typical examples, both using NLP on the Web, are Danuswan, Nishina, Akahori, and Shimizu (2001), who described the development and evaluation of a Thai learning system, and Nagata (2002), who developed *Banzai*, a Web-based language-learning application for learning Japanese grammar. A strong collection of work in this area was presented in the special issue of *CALICO* (Vol. 20, No. 3) called "Error Diagnosis and Error Correction in CALL," edited by Heift and Schulze (2003). Pujolà (2001) and Bangs (2003) gave excellent introductions to issues relating to feedback.

The advantage claimed for ILTS feedback over the more straightforward feedback given by conventional CALL programs is that more sophisticated error-specific feedback can be generated (see Toole & Heift, 2002). Errors can even be filtered and ordered so that feedback concerning the most significant errors is prioritized and presented first (Heift, 2003).

Toole and Heift (2002) offered a good example of an ILTS in their description of the *Tutor Assistant*, a Web-based authoring tool designed for learning English as a second language (ESL). In one version of this system, called the "ESL Tutor," the system analyzes sentences entered by students, detects grammatical and other kinds of errors, and then generates meaningful and interactive vocabulary and grammar practice. With regard to learner feedback, Toole and Heift (2002) noted:

> Unlike other ILTSs, feedback is also individualised through an adaptive Student Model, which monitors a user's performance over time across different grammatical constructs. This record of strengths and weaknesses is used to tailor feedback messages to learner expertise within a framework of guided discovery learning: a beginner student will receive the most explicit

feedback while the instructional messages for the advanced learner will merely hint at the error. (p. 375)

According to the authors, the *Tutor Assistant* is the first authoring system created for an ILTS. In this regard, Toole and Heift raised the very important issue of the time and expertise that language teachers require to produce new language-learning materials with an ILTS. The time commitment using an ILTS can be substantial. In contrast, there is no question that relatively simple authoring tools such as *Hot Potatoes* enable language teachers to produce exercises very quickly. Although the gain in having access to more sophisticated feedback mechanisms in an ILTS is important, so too is the teacher's valuable time. Ultimately, it is going to be a question of cost versus gain: whether the advantage of having a more intelligent system is enough to outweigh the disadvantage of having to spend more time authoring new materials. With well-conceived and careful design, Toole and Heift showed that time-efficient authoring of an ILTS is indeed possible.

Recently, ILTS and ICALL systems have been making increasing use of corpus and concordancing techniques. Cowan, Choi, and Kim (2003) illustrated how a large corpus of L2 learner errors can be very useful for identifying persistent grammatical problems. Having a large database or corpus of learner errors incorporated into the design of an ILTS—one that is used alongside more traditional grammars and dictionaries—enables even more sophisticated processes of error analysis to be achieved, so the nature and relative importance of recurrent L2 learner errors can be pinpointed even more precisely.

Design of CALL Tools

With CALL tutors, the software or system is designed to evaluate student input, which is then analyzed in order to generate feedback for the student. In contrast, with CALL tools, the computer is cast into a different role (see Levy, 1997, 2000). Instead of setting the computer into the role of tutor for human–computer interaction, with computer tools the role of the technology is best described as an "enabling" device. Thus, the tool might facilitate access to and act as a means of searching a database, or provide access to a communication technology and a software interface to facilitate its use (e.g., e-mail). Computer tools include searchable language databases, archives, monolingual and bilingual dictionaries, as well as the tools that fall under the rubric of computer-mediated communication (CMC; see chap. 4). CMC refers to communication between humans that is mediated by computer technology. It covers a wide range of synchronous (real-time) and asynchronous (delayed) forms of mediated communication, and includes e-mail, chat, text-based conferencing and videoconferencing, discussion lists, and mobile technologies. When these tools are used for the purposes of language learning, collectively we refer to them as *CMC-based CALL*.

For some years now, general-purpose tools have been used for language learning. E-mail, and to a lesser extent chat, have proved particularly popular and there are many examples of their use in the CALL literature (see Egbert & Hanson-Smith, 1999; Stockwell, 2003a; Tudini, 2003; Warschauer, 1996b). On the whole, when using these tools, designers have focused solely on the pedagogy and on creating appropriately designed tasks, rather than on considering how the technology side of the language-learning environment might be improved.

More recently, certain CALL designers have been shaping these general-purpose tools, both technically and pedagogically, so as to sharpen their focus and effectiveness for language learning. A good example is the work of Appel and Mullen in their development of tandem learning principles through the electronic tandem resources (ETR) Web site (see Appel & Mullen, 2000, 2002; see also Cziko, 2004).

In essence, e-mail tandem learning refers to organized language learning exchanges between two language learners who both want to improve their proficiency. According to Brammerts (1996), there are two fundamental principles of tandem learning: the principle of reciprocity and the principle of autonomy.

Thus, pairs of students are matched together so that each language learner is partnered with a native speaker of the target language, in a reciprocal way. The first principle requires each learner to contribute as equally as possible in each of their email exchanges. This means each learner must write half of their e-mail in their L1 and half in their L2. The second principle requires each learner to take responsibility for their role in the exchange.

In their paper, Appel and Mullen (2000) described the pedagogical advantages of their program over traditional e-mail tandem learning. Because students have password access only to their own account, they are not able to write messages to anyone other than their partner. As Appel and Mullen pointed out, "In this way, the environment is specifically oriented to the task of language learning, and concentration is focused, preventing the usage of this environment for other purposes which may be a distraction for the students and more properly accomplished through the use of their personal email accounts" (p. 294). From these beginnings, Appel continued to refine the ETR language-learning tool working through four iterations and steadily moving toward a more focused and refined product (see Appel & Mullen, 2002). In Version 4 there were a number of important new developments (Appel & Mullen, 2002):

1. The addition of a fully developed teacher interface, where teachers can monitor the work of their students without accessing the actual messages (privacy for the students is maintained), add exchange accounts, and communicate with students;

2. Changes in the student interface, to enhance usability and transparency, as well as an expansion of the automatic monitor function, all allowing students to concentrate more fully on the language task; and

3. Perhaps most importantly from a pedagogical standpoint, the addition of a class wide "Group Board" environment, where the teacher and all the students can post messages and communicate as a group. (p. 197)

Overall, the ETR resource contains sophisticated mechanisms for research purposes, including a tool for content analysis to indicate the balance of language use between L1 and L2 in learner exchanges, and a variety of tools to track the timing and effectiveness of the interactions.

The work of Appel and Mullen (2000) evolved out of a desire to better shape a general-purpose communication tool, for specific pedagogical purposes. A similar desire has led CALL designers at the Open University to better focus another CMC tool, audiographic conferencing, for language teaching and learning. Their work led to the development of a virtual learning environment (VLE) called *Lyceum*, developed for educational use by the Knowledge Media Institute at the Open University (see Hampel, 2003; Hampel & Hauck, 2004). Again, pedagogical decisions are very much in the forefront of the design. It includes a range of tools, among them "audio, concept map (developed for concept mapping, but also useful for making notes or brainstorming), whiteboard (for writing and drawing and for importing and manipulating Web images), text chat, and document module (for writing, discussing and editing longer texts)" (p. 22). The design of these increasingly sophisticated pedagogical tools emerging out of mainstream community technologies may be expected to continue. The only drawback is the significant time and resource commitment required for their continuing development and evaluation.

DISCUSSION

Introduction

Notable changes have occurred over the last several years as far as CALL design is concerned. Whereas once a CALL design might have been conceived within the bounds of a single theoretical or pedagogical orientation and a single technology solution, now the situation is increasingly more complex and designers draw on two or more complementary theoretical perspectives, often accompanied by two or more complementary technologies. There have been many examples described in this chapter. In addressing the projects and design processes in the first part of this chapter, certain topics and issues come to the fore. Principal among these is perhaps the multidimensional nature of the design process as developers try to weave together elements of theory, pedagogy, technology, and best practice, often drawn from a number of different fields or disciplines. By observing the different approaches taken, it is clear that CALL authors go about the business of design in different ways, drawing on a variety of knowledge bases and theoretical motivations, choosing from a range of starting points, and observing different priorities. All these aspects help shape and inform the final design. Although for some authors a course specification might be at the forefront of their thinking, for others a theory or theories might be the driving force, or yet again a particular choice of technology. Commentators often argue that technology should always be kept in the background, but the view taken here and by others in CALL (see Armitage & Bowerman, 2002; Kohn, 2001; Levy, 1997) is that technology concerns are implicated from the start, especially when we acknowledge that each delivery system or software development tool has its particular limitations as well as its particular strengths. It is only with a clear understanding of the strengths and limitations of different technologies that hybrid combinations can be created effectively, as we saw in the example of Strambi and Bouvet (2003), or that general-purpose technologies can be customized and improved, as we saw in the example of Appel and Mullen (2000).

Examples such as these leave us in no doubt that modern CALL design is becoming increasingly complex, and there are many decisions and choices to be made. What is common to all the projects reviewed here is the principled approach that the designers have employed. They each have taken great care to describe what they are doing and why, and detail is given. This is extremely reassuring. Although the discipline of CALL is undoubtedly becoming more complex, designers are making great efforts to meet the challenge by absorbing new developments and options, and building clear and convincing rationales for their work.

Also primary in the designs considered here are the authors' efforts to get to know their audience; that is, their needs, goals, and characteristics. The designer needs to make every attempt to get to know the potential users and the learning context. To quote Shneiderman, "[B]efore beginning a design, the characteristics of users and the situation must be precise and complete" (1987, p. 52). Without a clear understanding of the learner and context-related attributes, CALL designs will founder. Also, as recognized in many of the examples, design is fundamentally a creative process, which often involves the discovery of new goals. Shneiderman, who has been most influential in the area of interface design, and who has reflected profoundly on the development process, argued that we are not able to break the process down into separate, smaller units:

- Design is a *process*; it is not a state and cannot be adequately represented statically;
- The design process is *non-hierarchical*; it is neither strictly bottom-up nor strictly top-down;
- The process is *radically transformational*;
- Design intrinsically involves the *discovery of new goals*. (1987, p. 391)

Unless the development process is intentionally constrained—as in a controlled experimental study, for example—what is initially conceived may undergo significant, even radical, change as a result of the development process and feedback from users. Finally, the hardware and the software development tools that are employed exert a wide-ranging influence on design. All such tools have their strengths and limitations, especially in the interaction options that they offer, and these attributes variously shape and direct the design.

Integration: Horizontal and Vertical

As we saw in the last section, design is a complex and challenging activity because of the many—sometimes conflicting—ideas and elements that have to be made to work together successfully. The term used to describe this process and goal, of managing and drawing together the various components of a design, is called *integration* (see chap. 9). Integration has been a topic of interest through much of CALL's history (Appel & Mullen, 2002; Brussino, Luciano, & Gunn, 1999; Farrington, 1986; Garrett, 1991; Rogerson-Revell, 2003; Strambi & Bouvet, 2003; Trinder, 2003). Like design, integration can be viewed and understood in different ways. Three perspectives are introduced here in relation to design. Successful integration often requires an understanding of all three perspectives.

The Language Teacher's Perspective.   The first perspective is that of the language teacher. In this respect, integration often means thinking about ways of combining classroom and lab-based learning (if computers are not available in the classroom), or combining in-class and out-of-class work (e.g., homework), or the offline and online elements of a course. This perspective is often expressed in terms of how to best integrate CALL into the curriculum (and is considered at length in chap. 7). Integration of CALL activities into the curriculum can actually change the curriculum and add an important new dimension, but this can only happen if the curriculum is allowed to change.

Curricula themselves can vary along a continuum, from those entirely predetermined in advance to those that are very open ended and negotiable. Curricula at the predetermined end of the continuum can potentially

[handwritten: prior to this class, I have not viewed computers as a collaborative tool.]

*[handwritten: * integration into curriculum]*

provide both teachers and learners with clearly defined goals that can positively influence motivation. On the other hand, predefined curricula can be very restrictive for CALL because they simply do not allow sufficient time or space in the classroom for CALL activities to be introduced and integrated effectively. Often, with a resolutely fixed curriculum, CALL, if it is used at all, will tend to be a peripheral event and will lean toward the mundane and unimaginative.

If, on the other hand, the curriculum is designed in such a way that there are choices available to teachers and students, then more can be achieved with the CALL options that are available, both inside and outside the classroom. If taken too far, however, the limitation with a fully negotiable curriculum is that teachers and learners do not clearly perceive learning goals and objectives in the longer term, and learners feel they are not receiving the guidance they expect and deserve.

Our view is that a middle path is both possible and advisable, although this may well not suit some cultural and institutional contexts. The advantage of a middle path is a curriculum devised not as a prescription, but rather as a guide or template, where there is considerable room to move within it as far as the specific content is concerned. This approach, for instance, enables up-to-date authentic materials to be introduced from the Web, and can enable key-pal projects between native speakers (NS) and nonnative speakers (NNS). It also allows us to think about computers in the classroom as well as computers together in a laboratory; students working collaboratively with their peers; and teachers working with students at the computer, rather than students only working at the computer on their own.

If the organizing principle of the curriculum is thematic, functional, grammatical, lexical, or some combination of these elements, provided that the curriculum is flexible, supplementary CALL work can be drawn in successfully. Our own vision of the curriculum would be one of carefully circumscribed flexibility based on a detailed understanding of institutional, teacher, and learner goals and expectations.

As far as CALL and the curriculum is concerned, a further point should not be overlooked. One of the great potential strengths of CALL is the possibility of students learning on their own in the self-access center, library, public access areas, or even in the home at a convenient time and appropriate pace. If managed and integrated properly, such work can encourage the growth of learner independence and autonomy (see Blin, 1999, 2004). In such a way, the time on task is effectively extended with clear benefits in language learning. For this work to be successful, however, learner training and curriculum integration need to be managed very carefully. (Many of these core ideas resurface in the discussion sections of various chapters in this volume.)

The Language Learner's Perspective. The second perspective that helps us to understand more clearly the relationship between integration and design concerns the language learner. When viewed in this light, the topic of integration introduces questions of continuity across the curriculum, as perceived by the learner, and the prospect of specialist knowledge being required for distinct learning purposes or courses. It is not, we believe, beneficial for a student to develop an entirely idiosyncratic perception of technology use from one course to the next across an educational institution. Some continuity in learning tools across programs and courses is clearly advantageous from the student's point of view; however, achieving this continuity can be problematic when online courses and the like tend to be developed independently, without communication or discussion, by faculty in the disciplinary or subject domains. Nor, in the broader sense, is it wise to ignore the technologies students choose to use in the wider world. Again, continuity is advised, provided that pedagogical objectives are not compromised, because of the positive effects of familiarity and streamlining, especially with regard to the learner training required for effective participation. For the language teacher-designer, with an audience of increasingly technology-literate students whose level of technical knowledge and expertise is steadily growing, such questions are important.

As teachers and designers, therefore, we need to understand what can be safely assumed, and what cannot, so that we can employ different technologies at a level that is appropriate, and offer supplementary training when necessary. This kind of integration might be referred to as *horizontal integration*.

Horizontal integration is apparent when we consider integration from the students' point of view as they engage with technology in different contexts. For example, they will inevitably use new technologies in the wider world, for either work or relaxation. They may also use new technologies within an educational institution if they are studying, not only in their language-learning subjects but in other subjects as well. Knowledge of a student's experience in these non-CALL environments is important for CALL. The point is that each student does not enter the language-learning environment as a blank slate, in terms of his or her understanding of technology. Instead, each student enters the language class with a unique knowledge and experience of technology. Apart from the purely technical aspects—knowing how to use a word processor, for example, or how to conduct searches on the Internet—students will also enter the language class with a set of perceptions and expectations of technology use in education. Some students will welcome its use; others will not. Some will have wide-ranging and deep knowledge and experience; others will not. Many students will fall somewhere in between these extremes. As far as design is concerned, the language teacher or CALL designer needs to know about these predispositions, and be sensitive to patterns of individual variation. Then, the designer has to learn to work with this knowledge and provide, as far as possible, for individual learner preferences and characteristics.

Knowledge about the students' technical background, experience, and predispositions may lead to many insights. Such knowledge has the potential to lead to design decisions that create learning environments for students with a familiar "feel," and interfaces that are intuitive and easy to learn because they draw on the students' previous experience. Such an approach can save an immense amount of time in learner training. If learners already basically know how to use the software because of prior experience, much less time will need to be allocated in the language class to help the students learn how to make use of the software or technology. This is a strong argument for making creative use in CALL of generic application programs such as word processors, presentation tools, and e-mail applications. This is not to say that the applications and technologies used in language learning should necessarily be the same or similar to those used in other courses, but where continuity and consistency can be achieved without interfering with the quality of the learning experience, this should be a goal. There is no point in using specialized hardware and software just for the purposes of language learning if there is no good reason. On the other hand, there are distinct advantages to an integrated approach that attempts to draw into the language class valuable knowledge and experience from outside.

An Institutionwide Perspective.   The third perspective on integration involves the institution as a whole. Viewed from this perspective, integration may be understood in terms of a university's policy on teaching and learning, and the technology support and training that is available for technology-oriented development work. The third kind of integration might be termed *vertical integration*. To appreciate this perspective, more of an institutional orientation is needed as we consider how technology use in the language area or department fits—or does not fit—into the ways that technology is used institutionwide. This perspective brings into view the policies and support systems of the institution as a whole. Institutionwide policy in teaching and learning, the hardware and software applications purchased and supported (e.g., the university-approved LMS), the locations of the machines and access policy, and the relationship between teaching staff and technical support staff are all crucial issues that inevitably have an effect at the local level, even though they are conceived for the whole institution. If possible, local design requirements need to take full account of the wider environment, especially as far as hardware and software development tools are concerned. Likewise, the infrastructure of the wider environment needs to take full account of local needs. The continuity or congruence between the local environment and the wider environment will not only affect the ease with which the CALL materials are developed initially, but also help to ensure the longevity and ongoing support of the CALL materials once they are produced.

This issue was addressed in a paper by Barr and Gillespie (2003), who described the creation of a computer-based language-learning environment at the scale of the institution, in terms of the "human, technical and physical resources, communicative structures, information management and cultural contexts" (p. 68). It is noteworthy and laudable that Barr and Gillespie moved well beyond technology-related issues in this study. The authors described and compared the universities of Cambridge, Toronto, and Ulster in order to assess the pedagogical effectiveness of the three settings. The study made the comparison under the headings of learning and teaching resources, communication; pedagogical strategies; and infrastructural strategies. As far as we know, this was the first study of its kind.

The importance of integration was paramount throughout the study. This includes what Barr and Gillespie (2003) described as "infrastructural integration," covering computer and other resources and a well-coordinated operational infrastructure. It also includes integration at the local level, as it is more commonly understood in the CALL literature. In this study, even though the profile of each university was different (including the extent to which it was resourced), each environment was found to be functioning effectively. Barr and Gillespie (2003) asserted:

> Although each institution has integrated computer technology into language teaching and learning in different ways, a key element of each environment has been the establishment of a common computer-mediated infrastructure, enabling effective information dissemination, resource distribution, communication and teaching and learning. No single common infrastructure would be suitable in all three; however, in each case, it was found that the environments created were valuable, especially in integrating elements of the teaching and learning process that would normally have remained apart. (p. 68)

The authors then concluded that "adequate technical resources and a management that is keen to integrate computer technology into all aspects of university life is a key factor in their success" (p. 68).

Although these institutions were able to provide an effective environment, it should not be forgotten that such institutionwide decisions are not neutral. Certainly, they allow certain actions and options to be directly supported but, equally and most important, institutionwide decision making proceeds on the assumption that the technology and support requirements of all disciplines and fields within the university are basically the same. Unfortunately, in a number of important aspects they are not. For example, general-purpose tools tend to be developed with only the English language and native speakers of English in mind. Although the situation is slowly changing, the support for the scripts of languages other than English is lacking. Increasingly, in many institutions this means that local, domain-specific needs are not met.

Having said that, the provision and support of university-wide development tools and applications means that CALL designers do have access and support when the capabilities of the development tool meet their particular needs. If the general-purpose, university-wide development tool can be used to meet some of the CALL designers' requirements, then a special purpose tool (language specific, possibly) can be used to compensate for any deficiencies resulting in a hybrid design solution. To some extent, this was the approach taken by Strambi and Bouvet (2003) in using the general-purpose tool *WebCT* for certain components of their course development and delivery, and a CD for others.

Today, in many university contexts around the world, there is a clearly recognizable trend toward a narrow consistency in hardware and software applications. Whereas in the past this may have meant choosing a PC-style computer over a Mac, increasingly this strategy also applies to the software applications that are purchased and supported. Thus, we see *WebCT* or *BlackBoard* chosen as the university-wide software development tools, and we find that these are the only tools for which technical support and training is available. Motivated by the three "C"s—cost saving, centralization, and consolidation of technical expertise—fewer and fewer options are remaining for the designer, at least in the sense of a broader software development support base being available. The upshot is that the developer has fewer design choices that are supported.

With the limited time and resources that CALL designers have as individuals—or as members of small, often grant-supported teams—such factors emanating from the wider environment must be recognized and their impact acknowledged. This does not necessarily require compliance in all respects, but projects will be easier if, without compromising on key issues, goals and design decisions at the local level are in sympathy with the broader infrastructure. This is also important for the longer-term integration and success of the CALL materials that are developed.

Increasingly, it seems, designers are turning toward learning management systems (LMSs) such as *WebCT* and *BlackBoard* to deliver and manage their courses (see Arneil & Holmes, 2003; Godwin-Jones, 2003). This certainly has some advantages, but also presents a number of difficulties. The problem is that LMSs have limitations. Although they provide a general-purpose environment for the integration of a number of tools that assist with presenting course materials, communicating with and among students, and producing simple quizzes, they also have significant drawbacks that tend to guide the designer along a narrow path. In this regard, Arneil and Holmes (2003) provided a very useful discussion of the benefits and drawbacks of LMSs, and they compared them with other authoring options such as *Hot Potatoes*. Arneil and Holmes also discussed "hybrid exercises" (p. 64), where they looked at such techniques as the option of uploading client side pages, such as those created by *Hot Potatoes*, into an LMS system, such as *WebCT*. There are complications, however, as the functionality and capabilities of one system inevitably have to be reconciled with the other one.

In an earlier paper, Arneil and Holmes (1999) addressed design issues relating to their authoring tool, *Hot Potatoes*, in terms of "exercise design, the ability to customise and control the output for different browser versions, user-interface design, ancillary technology and technical support" (p. 12). This discussion remains of interest, even though there have been many developments since. Arneil and Holmes considered the ways in which the available technological infrastructure forms and shapes the design decisions. Notable here are specific browser requirements forcing certain choices. Extensive end-user testing is necessary, because of the differences between browsers and between subsequent versions of the same browser. Again, in exercise design Arneil and Homes discussed the limitations of the medium and the need for materials designers to work with them, and they noted the balance required between flexibility and practicality. Because they were designing for novice and sophisticated users, a consideration that leads to several levels of configuration is needed if users are to operate at the right skill level. Arneil and Holmes concluded, "This complex mix of decisions and balancing acts underlies the design of all authoring tools ..." (p. 18).

In CALL, problems of design are always closely linked to the precise limitations of the technology applications in use. As a result of these limitations, certain trade-offs have to be made with the design. Once the technological constraints are known and understood, the researcher or developer can try to establish how these facts might shape or ameliorate the pedagogical ambitions. Understanding the limitations is crucial for effective design. User needs, abilities, and expectations may also function as a design constraint. This was evident in the discussion by Arneil and Holmes (1999), in which they were endeavoring to meet the needs of different authors and their skill levels on a variety of browser platforms. The point is that all new technology-mediated systems have constraints of one kind or another. CALL design requires us to understand exactly what they are, and the specific constraints they impose, and then to design effectively within these limitations. From the designer's point of view, CDs, LMSs such as *WebCT* and *BlackBoard*, and authoring tools such as *Hot Potatoes* all have their strengths and limitations (sometimes referred to as *affordances*—see Gibson, 1979; Hutchby, 2001) that both enable and constrain a design. New options need to be evaluated in a similar way. Also, it is important to keep an eye on authoring tools for ITS, such as the one described by Toole and Heift (2002), because of the more sophisticated feedback mechanisms these systems can provide, even in simple interactive exercises.

The argument for some agreement between CALL developers and the technological and policy infrastructure of which they are part is largely motivated by a desire to see CALL materials have a longer life span. This is a crucial consideration given the time and effort involved in material creation and is an issue that has not received sufficient attention thus far.

Improved vertical integration has the potential to contribute very positively in this respect, as has the adoption of a more integrated approach generally, from having technology easily accessible in the classroom to ensuring that there is continuity, whenever possible, between the applications that learners use in their language study and other subjects, and at home and at school, for example. Armitage and Bowerman (2002) tackled this problem too, although from a more technical angle. In their article, they discussed "knowledge pooling" and made some valuable points about actions that may be taken in design to help achieve greater longevity for CALL materials. They argued for the "need to build systems which allow effective reuse and maintenance of their components, and offer the opportunity to add new components; and second, the need to build systems which keep the data and design strictly separate, as this is the foundation upon which reuse and scalability are raised" (pp. 27–28).

The idea of integration, or an integrated learning environment, is a very important one for CALL. This idea is returned to at points throughout the book, with an extended discussion in chapter 9.

A Focus on the Learner

We saw earlier in relation to the discussion on horizontal integration that an understanding of the learners' background, role, and perspective is crucial in design. This understanding has been well accepted and represented in CALL, especially by authors such as Dominique Hémard. Through a series of research and development articles over a number of years, Hémard has emphasized the importance of a user-centered approach to the design process (Hémard, 1997, 1999, 2003; Hémard & Cushion, 2002; see also Shin & Wastell, 2001). He has maintained that a principled and rigorous approach to design is still lacking in much CALL design practice, and has argued that "design is still too often driven by its perceived technological potential whilst being all too clearly affected by its own limitations" (Hémard, 2003, p. 22).

The approach to the design process that Hémard (2003) advocated is a well-proven design methodology drawn from human–computer interaction (HCI) studies. Hémard described the interdisciplinary field of HCI:

"Its main tenets are that a system is more than the software alone, that its scope and purpose are wider than its functionality and that the larger system, including the human users and the physical, organisational and social environments, must be considered in order to make appropriate decisions" (Preece et al., 1994)" (p. 22). Thus, the HCI approach to design is entirely compatible with the ideas presented in the previous section on integration. It is also compatible with a systems approach to design that argues that the elements of a system can only be understood if they are viewed as part of the whole, and that the parts of the system are dynamically interrelated so that a change in one part will exert an effect on the other parts (see Levy, 1997). This kind of thinking helps us understand how a university-wide decision (e.g., the decision to enforce a hardware upgrade, or to cease supporting a particular software application) has significant ramifications for users and developers working locally, especially if there has been little or no communication about the changes taking place.

Hémard (2003) presented a best-practice model for designing online CALL materials. This model provides a rigorous, user-centered approach that helps the developer through the design process. In brief, it seeks to help the designer define more specifically the "design" or "problem" space. Then, through active data collection procedures, it progressively seeks to identify and clarify user needs and goals. A variety of specialized HCI methods and techniques is used to create a design framework for a conceptual model and to define mental models that lead to a more precise understanding of user requirements. These are turned into specific design decisions to form concrete goals in the design of the new application. Finally, evaluation occurs and, through the use of guidelines and checklists, the design is refined in an iterative manner (see Hémard, 2003, for a detailed understanding of the approach).

For those who are new to CALL design, it is important to understand that these special techniques and approaches exist to help the designer. Given the breadth of knowledge required when designing new CALL materials, and the multidisciplinary nature of the task, it is not surprising that designers whose principal interest is usually language teaching and learning are not aware of HCI approaches. Yet, these approaches can be enormously helpful in bringing rigor and principles to the design process.

To prepare a strong and workable design, the designer must be sensitive to individual learner characteristics and the learning context. Ideally, designs would differ according to the conditions and parameters set by the learner and the learning context. The problem for designs grounded in a specific theory of language learning, or for commercially produced materials aimed at the wider market, is the designer's lack of detailed knowledge about the individual learner and the curricular and institutional context in which the program will be used (see Shaughnessy, 2003). Specific knowledge of the learner's age, gender, physical abilities, education, cultural or ethnic background, motivation, goals, personality, computing experience, and so on remain largely unknown. In these circumstances, the designer may rely on intelligent guesswork or traditional responses to design problems. Moreover, the designer often attempts to cover all the bases, and here Shneiderman's (1987) observation applies: "If implementers find that another command can be added, the designer is often tempted to include the command in the hope that some users will find it helpful" (p. 55). The audience remains remote to the author, at least in any specific sense, and generalized predictions have to be made; arguably, this is easier at beginner level than at any other. There is no doubt that the better one knows the needs and characteristics of the user, the better one can make the design.

Wherever possible, to ensure that their needs are met, learners need to be involved in the design process. Learners can play an important role if they are given the opportunity. For example, students can write *Hot Potatoes* items for each other, or they can develop learning and review strategies to share with one another (P. Hubbard, personal communication, 2005). Another telling example of how students might be involved was described by Crompton (1999) in the development of a CALL grammar course.

An important feature in the design of Crompton's (1999) course in grammar for learners of Spanish was the student progress log. Interestingly, even though the students' progress logs were carefully managed and maintained online, the students *insisted* on "handing in" a hard copy of the progress log. Crompton noted of the importance for students of "the ritual of handing something in" and, relatedly, the teaching and learning process as "an extremely complex *anthropological* activity" (1999, p. 79; italics in the original). The students' need for a visible record of their progress prompted an unexpected move back to a hard-copy student progress log in Mark II of the course. Another critical refinement came about as a result of recording the students' log-in times and scores in the exercises. Surprisingly, there was an inverse ratio between the length of the exercise and the amount of time spent completing it; in other words, the longer the exercise, the shorter the time the student was logged onto it, and vice versa.. This appeared to be due to students leaving the exercise for a time and "surfing" on the Internet. In this regard, Crompton made some crucial points about design:

> [A]s well as short exercises there has to be a variety of activities within the units to keep the attention. The Internet is not electronic text; it is a much more exciting place, and any materials placed on it have to reflect that excitement, above all in the range and variety of activities offered. Just as no-one would think of producing a videotape consisting entirely of scrolled text, so the Internet has to match the professionalism and variety of other websites. It has to be *interesting*. (p. 80; italics in the original)

These and other considerations drawn from student feedback and user monitoring led to a much refined model for Version II. The refinements included:

- shorter exercises;
- a variety of hyperlinks with each exercise;
- inclusion of realia to enable exercises with "real" links;
- the whole module can be downloaded for home use;
- discrete test;
- external accreditation. (Crompton, 1999, p. 81)

Not surprisingly, Crompton concluded that the Mark II version of the course is "quite a different type of Internet activity" and certainly a much more effective one (p. 82). In sum, although it is undoubtedly a challenge to match the flair and quality of the better commercial Web sites with the resources we have as educationalists, we have to appreciate that the online experiences students have in the wider world will contribute to be standards that learners expect when judging our CALL materials. The problem is not insurmountable, but it does require us to be thoughtful and creative designers.

CONCLUSION

In this chapter, we have looked at a number of recent projects in detail and considered many topics and issues related to design in contemporary CALL. From this discussion, we would suggest a number of guidelines for designers:

1. Know the strengths and limitations of existing CALL materials.
2. Be aware of your technology infrastructure, including your institution's favored LMS and the technical support available.
3. Know your prospective audience, as a group and as individuals.
4. Know the strengths and limitations of your development tools.
5. Review possible ways of approaching design problems, such as,
   a. Point(s) of departure
   b. Theory base(s)
   c. HCI principles.
6. Incrementally refine the project goals and the design space.
7. Where possible, link theoretical elements directly with specific design features.
8. Test, retest, and evaluate with users.
9. Be creative.

To close the chapter, we would like to quote from the work of Kurt Kohn (2001). We think this summarizes the current circumstances for CALL designers perfectly:

> Those who approach the future of technology-enhanced language learning (TELL) from within its human and pedagogic heart know only too well that what we are witnessing today is a dramatic process of diversification and complexification. The range of available learning and teaching options will be richer and they will (have to) be capable of seamless combination depending on needs, preferences, feasibility, and pedagogic wisdom. Teachers who set out to facilitate learning processes in technology-enhanced training environments will need pedagogic and linguistic competencies that are clearly beyond those that were sufficient in the past. Overall, the task is getting more complex, more demanding and more rewarding. (p. 252)

Although the chapters that follow this one explore other dimensions of CALL, the topic of design keeps reappearing. Design is fundamental to the field. It is the tangible means by which theory and research findings are made explicit and put into practice in the structuring of CALL tasks, programs, and online courses.

Chapter 3

Evaluation

It is appropriate that the evaluation chapter follows the design chapter, because of their strong connection and the frequent overlap between them. Most language teachers and teacher-designers using CALL want to evaluate their work, if not at points during the design process then at least at the end of the project. Typically, language teachers want to be able to assess student attitudes and perceptions in a learning environment that involves technology; they want to know whether CALL tasks are working as they should; and they are interested in the viability and effectiveness of specific methodologies and strategies, often with a view to refinement and improvement in practice. The focus in many evaluative studies is on assessing the value and effectiveness of the CALL materials that are created. These materials include software, Web sites, online courses, computer-mediated communication tools used for language learning, and more complex, multifunctional environments involving a combination of elements, such as the LMSs described in the last chapter. It is evident from this short list that the targets for evaluation in CALL are many and various.

In going about their evaluative work, CALL practitioners use a wide range of methodologies, from the simple to the sophisticated. At one end of the scale is the simple checklist or survey. At the more complex end of the scale are multifaceted, longitudinal evaluation studies that may involve qualitative and quantitative approaches (e.g. Harless et al., 1999; Trinder 2003). For example, Harless et al. (1999), in their evaluation of a multimedia method using virtual dialogues with native speakers, conducted three kinds of evaluation studies that complemented one another: a technological feasibility study; a qualitative, phenomenological study; and a quantitative experiment with a pretest, posttest, survey, and interview. In addition, the well-known frameworks of Hubbard (1987, 1988, 1992, 1996) and

40

Chapelle (2001) offer a more sophisticated foundation for a principled approach to evaluation in CALL.

CALL evaluations can also be large scale, as when Barr and Gillespie (2003) looked at pedagogical effectiveness at three universities, as described in the last chapter. Evaluations can even take place on a national scale, such as the web-enhanced language learning (WELL) evaluation project in the United Kingdom (see Haworth & Cowling, 1999). But for the most part in this chapter, and in relation to CALL, we look at evaluations on a smaller scale.

Given that this chapter focuses on evaluation, and chapter 6 focuses on research, it is important that the difference between evaluation and research is clear, as it is understood here. To make the distinction, we have drawn chiefly on Krathwohl (1993) and Neuman (2003), who discussed the methods of educational and social science research, and on Johnson (1992) and Chapelle (2001) in the area of second language learning research.

Johnson (1992) provided a valuable overview statement on the key differences when she asserted, "The purpose of an *evaluation* study is to assess the quality, effectiveness or general value of a program or other entity. The purpose of *research* is to contribute to the body of scholarly knowledge about a topic or to contribute to theory" (p. 192). However, Johnson immediately explained that the distinction is not as clear-cut as it appears and there is overlap, as one finds in CALL. Still, the distinction is a useful one as long as it is not considered absolute. In CALL, the objects of evaluation studies are typically online courses, programs, Web sites, tasks, and learning environments. In evaluation studies, as opposed to research studies, the focus is relatively narrow and contained. Krathwohl (1993) distinguished evaluation from research not by its methods, which are usually drawn from the methods of research, but in other ways. He maintained that the key difference is that evaluation is *decision driven*. This stands in contrast to research that is motivated by hypotheses or more open-ended research questions. Also, with regard to the evaluation–research nexus, it should be noted that in many cases an evaluator does not want to conduct research: He or she may not have the time or the interest, and clearly research does not necessarily have to be the end goal. Instead, the evaluator may simply require an effective and efficient way to reach an informed and well-reasoned decision on an aspect of practice using a clearly articulated process that can be reliably followed. Often, for language teachers or CALL materials designers, their goal may be a personal one, and the evaluation process offers a way to be reasonably well satisfied that an approach or design feature is working as well as it should.

Hémard and Cushion (2001) provided a good example of a decision-driven evaluation. Their 12-month evaluation of a Web-based CALL project is decision driven in that the results of the evaluation led to a series of specific changes to the design of the program interface and the authoring platform. These changes concerned "the display, the functionality, the user interaction as well as the feedback and the help provided, in an attempt to establish a closer match between requirements and design" (Hémard & Cushion, 2001, p. 29). Evaluation led to informed decision making and, in this case, not to a single decision but rather a series of related decisions about the program's design. More broadly, Krathwohl (1993) argued that the main goal of an evaluation is to reach a decision on "the worth of something" (p. 524); or to learn whether a program or activity has met its intended objectives (Neuman, 2003).

In addition, the audience for evaluation and research studies differs in that evaluations are generally directed toward "a more targeted group than research," often referred to as the "stakeholders" (Krathwohl, 1993, p. 524). A further observation made by both Neuman and Krathwohl is that the value of an evaluation may often lie in its usefulness in a process; the evaluation process may be as important as the product. Neuman (2003) added that evaluation studies tend not to be theoretically oriented, although he quoted Brickmayer (2000) who supported theory-based evaluation that moves beyond "Did it work?" questions to "Why did it work or not work?" questions. That said, for the most part this chapter restricts itself to evaluation studies of the "Did it work?" kind, whereas chapter 6 on research deals with studies that aim at a deeper, more explanatory analysis.

In summary, then, as understood in this chapter, evaluation studies:

- Are aimed at establishing the worth of something.
- Are primarily decision driven.
- Are designed for a more narrowly defined audience than are research studies.
- Have a practical outcome.
- Draw value from the process as well as from the product of the evaluation.
- Focus on "Did it work?" questions rather than "Why did it work?" questions.

That evaluations draw value from the process as well as from the product means that evaluation methodologies are a central topic in this chapter.

Before continuing, we should also note at this stage the basic distinction between formative and summative evaluation. In the context of computer-assisted learning, formative evaluation is conducted during the development of new materials, with the goal of strengthening or improving them. In contrast, summative evaluation occurs after completion of a project, or at a significant staging post within it. Thus, summative evaluation is applied to the project as a whole, or a significant, self-contained part of it. The two perspectives are important for a number of reasons. Generally, in formative evaluation, the questions asked are determined by the developers and the feedback received flows directly back into improving the project as it continues (e.g., Gimeno-Sanz, 2002). In contrast, summative evaluation may well be handled by an independent third party; the results of this kind of evaluation are as likely to serve the needs of external stakeholders as those of the developers themselves (see Söntgens, 2001).

This chapter traverses major issues in evaluation by considering the objects of evaluations in CALL and the corresponding criteria and methodologies adopted. This includes the use of checklists and surveys for evaluation purposes, designer-oriented evaluations that are usually more narrowly conceived, as well as larger-scale, more general evaluation frameworks and their function and role. The choice of evaluation criteria in relation to both the intended audience and the technology provides a major and recurring theme throughout. Otherwise, essentially, the discussion moves from the general to the specific, and the more straightforward to the more sophisticated and specialized approaches to evaluation in CALL.

DESCRIPTION

Approaches to Software Evaluation

The construction and testing of new materials or artifacts is one of the unique, defining features of CALL. In this field, construction and testing is as much a focus as are research and practice conceived around ready-made products or generic applications such as e-mail, Web browsers, or the word processor. The introduction of microcomputers in the late 1970s and 1980s enabled language teachers to create CALL software for the first time, rather than waiting for an invitation to be the 'language expert' in a software development team developing materials for mainframe computers. Artifact building can involve making materials using more established technologies such as CDs or Web sites, or newly emerging ones such as automated speech recognition (ASR). Those that deal specifically with an emerging technology that has not reached critical mass, such as the group evaluating ASR (see Holland, 1999), typically want to assess the viability of the technology in a CALL context. Thus, Eskenazi (1999) examined "speech interactive CALL," or "ASR-based CALL," and posed the research question, "Has the technology come far enough for systems to be able to teach pronunciation effectively?" (p. 447). These are legitimate questions; they need to be asked if the language teaching community is not to be continually surprised by a new technology once it reaches critical mass and very quickly spreads to the population as a whole.

As a result of these developments, the language teacher or designer is suddenly in the position of evaluator too, with the need to test the newly created software with students to see if it worked successfully. With the advent of the microcomputer, computer labs became increasingly common in language teaching institutions, and commercially produced CALL software became more widely available. Again, for the first time, language teachers and administrators had to make decisions on what software to purchase for their institutions, and they needed some kind of basis for doing so.

Checklists and Surveys. Perhaps the most commonly used form of evaluation in this regard is the evaluation checklist. The checklist has been a feature in CALL evaluation from the earliest days to the present (Hémard, 2003; Hope, Taylor, & Pusack, 1984; Hubbard, 1987; Jones & Fortescue, 1987; Knowles, 1992; Levy & Farrugia, 1988; Susser, 2001). The checklist by Levy and Farrugia (1988) provides a representative example. It was broken down into a series of categories with a set of questions for each category. This checklist, similar to many others of the time, contained the following 14 sections: content, program objectives, documentation, program instructions, student use, program response to student, program design, technical quality, authoring material, motivational devices, teacher utility, multiple-choice questions, difficulties for ESL students, and presentation. Each category included a series of related questions. For example, there were four questions under "Program Objectives":

1. Are program objectives commensurate with those of the college, the teachers and the students?
2. Are program objectives clearly defined in the documentation?
3. Are the stated objectives achieved?
4. Are the objectives relevant to the student? Are they clear to the student? (p. 57)

Such a list of questions was typical of these kinds of checklists. Often, as in this case, it was not made clear how exactly the evaluator should go about answering the questions, nor what the evaluator should do if some answers were positive and some negative. In other words, there was no methodology to accompany the checklist that might have given guidance on how the evaluator was to resolve issues when both strengths and limitations were indicated.

Perhaps the most common evaluation instrument after the simple checklist is the survey (questionnaire/interview), and it is used frequently in CALL evaluation. Examples of recent surveys are given in Table 3.1. These examples demonstrate the role of the survey as one of the ways of evaluating the design and development of CALL materials. In these examples, the artifacts or products included CALL programs, CDs, Web sites, and courses. In all cases, the survey provided a useful instrument to collect student and

TABLE 3.1

Goals of Evaluations Involving a Survey

Goal or Purpose of Evaluation	Author(s)
To evaluate a new technology, functionality, or application (e.g., automated speech recognition, mobile technologies)	Harless et al. (1999), Holland, Kaplan, and Sabol (1999), Thornton and Houser (2002)
To assess student attitudes and perceptions (e.g., toward CALL programs, CDs, Web sites, Web-based projects)	Gimeno-Sanz (2002), Hémard and Cushion (2001), Trinder (2003)
To obtain feedback from students and/or tutors on a CALL course (Web based), or courseware	Iskold (2003), Soboleva and Tronenko (2002)
To investigate learners' views on the feedback features of a distance-learning teacher-training course (e.g., technical and design features, relevance, possible pedagogical implications, method of delivery)	Ypsilandis (2002)

teacher reactions. The evaluation study by Soboleva and Tronenko (2002) was instructive. These CALL designers assessed the effectiveness of a CALL course they created called "Learning Russian on the Web" by observing and questioning students and interviewing teachers. The responses to the surveys revealed both the strengths and weaknesses of the online course. What is appealing in this example is the focused nature of the questions, the detail that was extracted from the surveys, and the way the course designers responded to the observations and concerns of those involved when making further additions and modifications to the course.

The checklist or the survey is often used for evaluation purposes by language teachers-designers who create CALL materials for themselves. Often, such instruments are employed to gain insight into how well a particular program, Web site, or online course is working, often with a view to later improvement. Evaluators in this situation, working in a known and specific context, will always be thinking about their own approach, students, and curriculum; they have the great advantage of knowing their students' background and needs well, a critical aspect in evaluation. This is not always the case, however, and there is another important category of evaluation approach to consider. Evaluations may also be conducted by those quite removed from the particular student group with whom the material is going to be used; evaluators may also be entirely removed from the design of the software, Web site, online course, or program. We have labeled this group "third-party evaluations," and these are considered next.

Third-Party Evaluations

Third-party evaluations arise when the evaluator(s) has had no direct involvement in the object of the evaluation. Thus, language teachers evaluating a new CALL package that has arrived at their school or university fall into this category. Similarly, an experienced practitioner who is writing a software review for a CALL journal is a third-party evaluator. In both cases, the person evaluating the item in focus has not been directly involved in its development.

For the third-party evaluator, the two main challenges involve choosing the appropriate evaluation criteria and really getting to know the software, not only in itself but in judging how best it might be used in different settings with different learners. Although the language teacher may know the specifics of the context and the students for whom the materials are intended—a distinct advantage for the language teacher over the software reviewer—if the evaluator has not been involved in the design and development process, understanding the potentials of the software and possible tasks that may be created around it can be a challenge. For the software reviewer, evaluation is even more of a challenge because particular student needs and characteristics have to be imagined or projected, and based on the evaluator's prior experience, and judgments need to be made on the appropriateness of the CALL materials for different settings and kinds of student group.

Getting to know a hypertext document like a Web site, where many different paths may be taken through the materials, is not easy, especially when compared with, say, a textbook, which is presented linearly and for which the size and scope are reasonably self-evident from the start. Not having been intimately involved in the design and creation of the materials can make it more difficult for the third-party evaluator to gauge the purpose and the potential of the materials. Additionally, for the software reviewer, the particularities of the audience are unknown.

All the same, software reviews collectively form an important resource. Software reviews can be very informative whether distributed locally, published on the Web, or, a little more formally, in the software review section of a CALL journal. Software reviews are also potentially very useful to language teachers, not only in what they say about the specific software but also in the way the better ones are constructed and in the principles that shape them. An understanding of these principles can be very useful in helping teachers when they conduct their own evaluations, either formally or more informally. Still, the disadvantage is that the software reviewer does not know the students in the way that a classroom language teacher conducting an evaluation might know them. Thus, in some ways the task of the software reviewer is more difficult than that of the language teacher who is conducting a class-oriented evaluation.

Third-party evaluations are considered in this section by looking at an example of the principles underpinning the structure and content of software reviews in a leading CALL journal. A particular review using the template is then considered to illustrate how the framework operates in practice and how it leads to a decision on the worth of the materials; the particular example selected is a CD for teaching pronunciation. This is followed by a description of a leading Web site for English language learning to illustrate the key factors that need to be appraised when considering evaluation of this type of language-learning material.

The CALICO Software Review.   As it is a frequent source of software reviews, the approach taken by the *CALICO Journal* is a good place to begin when setting out to describe different approaches to evaluation. Burston (2003b) described the approach and set out the structure and style of the *CALICO Journal* software review. Rather than operating as a checklist, Burston noted that the approach looks at "critical systematic properties" using "an intrinsically discursive process" (p. 35). In general terms Burston stipulated:

> Before any software can be recommended for purchase, it must meet the first two of the following requirements and some combination of the last three:
> 1. Pedagogical validity
> 2. Curriculum adaptability
> 3. Efficiency
> 4. Effectiveness
> 5. Pedagogical innovation (p. 35)

Then, based on Hubbard's framework (see detailed description later in this chapter), the following four categories are used to create the basic template for software evaluation in the journal:

1. Technical features.
2. Activities (procedure).
3. Teacher fit (approach).
4. Learner fit (design).

The first category covers the basic features one would expect to find, such as the reliability of operation, ease of use, and so forth, and the second category also is relatively straightforward in covering the nature and design of the activities. The third and fourth categories are a little more intricate. Burston (2003b) stated:

> An assessment of teacher fit primarily involves looking at the theoretical underpinnings of student activities; judging how well they conform to accepted

theories of cognitive development, second language acquisition, and class-room methodology; and determining how closely they accord with the teacher's curricula objectives. (p. 38)

According to Burston (2003b), *teacher fit* is the most critical component of software evaluation, and the most difficult to assess. Finally, with *learner fit*, he asserted, "[T]eachers are in essence defining the potential user of the software program. In doing so, they are also determining the extent to which the program is appropriate for, or can be adapted to, the needs of their students" (p. 39). Here, learner fit includes linguistic level, response handling, adaptation to individual learner differences, learning styles, learning strategies, learner control, and design flexibility by the instructor.

CD: Connected Speech.   A software review from the journal is consid-ered briefly now as an example of the approach taken in the *CALICO Jour-nal*. The program concerned is called *Connected Speech* and was reviewed in the journal by Mark Darhower (Darhower, 2003). This product is an inter-active multimedia computer program presented on a CD for teaching pro-nunciation and effective communication skills to ESL and EFL learners. More specifically, with regard to the goals of the program, in the accompa-nying booklet the program designers asserted:

> *Connected Speech* is designed to assist the learner to identify, to understand the importance of, and to be able to produce the suprasegmental features of spoken English. It is well-known that suprasegmental features, such as word stress, centred stress and pitch change, play a major role in effective commu-nication.

For our purposes in this chapter, the goal is not to report on the outcome of the software evaluation, nor to repeat the details of the review. Rather, some of the features of the review are highlighted, especially where they contributed to an understanding of the evaluative process and the approach taken.

In terms of format, the structure of the evaluation section of the review was basically that described previously for the *CALICO* software review. Be-fore the main evaluative section there were short additional sections called Product at a Glance, giving basic information and technical requirements, and General Description, giving a summary of the features, background in-formation, and the documentation available. The evaluation was presented in the four sections, as described earlier.

Given the central requirement in this program for accurate voice recog-nition and, a little less centrally, for effective feedback, it is not surprising that these capabilities of *Connected Speech* were highlighted in the review. This relates strongly to Burston's (2003a) "critical systematic properties" mentioned earlier. In this evaluation, the critical property of reliable voice recognition was not difficult to identify, although it may have been difficult to assess accurately given the variability in microphones and so forth from location to location. In other CALL programs, the critical systematic properties may be harder to identify. In any case, the reviewer needed considerable expertise in relation to the critical property, in order to evaluate it fairly and at an appropriate level of analysis. In this review, much of this discussion on the accuracy of the voice recognition function was located in the Technology Features section. In the Activities section, the content structure of the program as a whole is described, and, within that, the different kinds of exercises were discussed. Special features were reviewed, such as the learning mode and the test mode for each activity, and the way in which the visual and audio feedback works when responding to users' correct and incorrect pronunciation. Here, the reviewer has clearly tested the many elements of the program thoroughly, and any shortcomings were noted.

In teacher fit, Darhower (2003) looked at the assumptions underpinning the design of the program and pointed out the central design feature, identified by the authors, that suprasegmental features were most effectively presented within extended samples of natural speech. As such, a communicative context for learning the phonological features of the language was provided. In this section, the reviewer also considered the sociocultural content of the program and the linguistic accuracy. Again, for this reviewer, these aspects came to mind as significant features of the approach, and were worthy of comment. Finally, in learner fit, Darhower considered the program from five perspectives: linguistic level, response handling/feedback, learning styles and strategies, learner control, and design flexibility. The reviewer was making informed choices appropriate to the program to organize his commentary, having chosen five of the seven properties potentially affecting learner fit described by Burston (2003a). In a summary section at the end, Darhower wrapped up the review by giving an overall evaluative statement on the quality of the program and an answer to the fundamental question, "Does it work?"

The CD format is particularly suited to the approach taken in the *CALICO Journal*, as discussed later in this chapter, and the key evaluation criteria are clearly evident in this review of *Connected Speech*. In the next example, we move away a little from this kind of CALL artifact or product and consider some of the issues that arise when trying to formulate a principled approach to the evaluation of a multifaceted language-learning Web site.

Web site: Dave's ESL Café. *Dave's ESL Café*, created by Dave Sperling and first released in 1995, is perhaps one of the best-known Web sites for learning EFL/ESL. Unlike some Web sites, it has been around for some time now and has achieved a degree of maturity and stability in its offerings. This Web site is a composite entity and contains many elements; this combination of elements creates more of a challenge for the evaluator, and this characteristic is one of the reasons it was selected as a focus here. The site contains not only a wide range of language-learning materials and activities, but also, through the communication tools it provides, a team of teachers around the world who help answer questions in the Help Center, described as a global virtual classroom for ESL/EFL teachers. Hence, it is a site that not contains only materials, but also, in a very important sense, people. As such, it represents a forum for language-learning activity in addition to any static content or exercises it might provide.

The wide range of resources that are hyperlinked to the homepage gives an indication of the diversity of *Dave's ESL Café*. The main links are given in Table 3.2.

This site is very rich in terms of the varieties of activity available. Under Hint-of-the-Day, hints on grammar and the relationship between English sounds and spelling are provided, usually in the form of rules with examples. Phrasal Verbs and Slang are presented similarly, with definitions and examples. The Quizzes section includes questions on many topics including geography, grammar, history, idioms, reading, and writing. In the Quizzes section, language learners are given a gap-fill sentence with two or three alternative answers. After completing 10 sentences, the students submit their answers and the system provides feedback, either right or wrong, for each question. This kind of feedback is the simplest available (see Bangs, 2003; Heift, 2003; Pujolà, 2001; Pusack, 1983). Chat Central, the Discussion Forums (Teacher Forums and Student Forums), and the Help Center are all examples of computer-mediated communication (CMC) tools that facilitate interactive exchanges of different kinds between users.

TABLE 3.2

Main Links on the Homepage of Dave's ESL Café

Stuff for Teachers	*Stuff for Students*	*Stuff for Everyone*
Idea Cookbook	Help Center	Chat Central
International Job Forums	Hint-of-the-Day	Photo Gallery
International Job Board	Idioms	Quotes
Job Center	Phrasal Verbs	Today in History
Job Information Journal	Pronunciation Power	Web Links
Korean Job Forums	Quizzes	
Korean Job Board	Slang	
Teacher Forums	Student Forums	

This kind of Web site contains a range of activities that differ markedly in the nature and quality of the interactivity they provide. There are basically three types of item in the Web site in terms of interactivity. Some of the elements in the site (e.g., Hint-of-the-Day) are simply descriptive texts, so the interaction, if it can be called that, is unidirectional (Type I). Another group of activities (e.g., Quizzes) are a little richer in interactional terms. They are corrected by computer tutors using a simple matching algorithm; the interaction is two-way and rudimentary feedback (e.g., right/wrong) is given in response to student input (Type II). Yet other activities are more sophisticated again (e.g., the Help Center); here, online communication tools are used to enable contact and interaction between students, and between teachers and students (Type III).

These qualitatively different kinds of interaction lead us to consider carefully our evaluation criteria. In Type II and Type III interactivity, the question of whether humans or machines are giving the feedback is all important. The computer giving simple right/wrong feedback on multiple-choice questions is operating at an entirely different level of sophistication compared to the human teacher who is responding to a student enquiry. In the Help Center, for example, students can ask a wide range of questions on a wide range of topics, including grammar, writing, and vocabulary. In one instance, a student sent in a request for the verbs in a paragraph to be checked and corrected. The human respondent generously went through the paragraph indicating errors with capital letters. (Posted by Jim, in response to: *verb corrections* (Eve)—8 February 2005, 9:40 a.m.) In such cases, correct feedback is provided for individual student questions. The quality of any human response or feedback is typically well ahead of what a computer can provide, because of a human's ability to grasp the significance of contextual factors. Consequently, one-to-one teacher– student interactions are likely to lead to thoughtful and accurate feedback, and impressive evaluation results. Clearly, sites that contain all three interactivity types might rate very highly in some areas (Type III) and rather poorly in others (Type I or II). It is therefore difficult to give a meaningful global rating for the site as a whole.

Evaluating such multifunction sites can be time consuming because each component requires consideration. The decision-making aspect amounts to an activity-by-activity assessment based on appropriately chosen evaluation criteria for each item. Some options/activities are likely to prove useful, whereas others may not. Hence, here the "Does it work?" question should not be applied to the entire Web site as if it were an undifferentiated, single entity. Instead, the various activities need to be considered one by one. Thus, the result of this process may be insights into the nature of the interactivity in each case and specific decisions on which activities are most appropriate for a particular group of students.

A number of other language-learning Web sites follow the same pattern in providing a wide range of exercises and activities. One expertly designed site in this respect is *Learn Welsh*, mentioned in the introduction, a real benchmark in terms of variety and all-round quality (e.g., http://www.bbc.co.uk/wales/learnwelsh/). Further examples include many well-designed pay Web sites, such as *English Town* (http://www.englishtown.com/), which is organized around language-learner proficiency levels with, again, many parts and sections that would need to be considered individually for evaluation purposes because they are so varied and different. However, it must also be said that not all language-learning Web sites fall into this category and offer such a variety and range of interactivity types. Some sites only provide a single kind of activity, or focus on a particular language area or skill, such as *Randall's ESL Cyber Listening Lab* (http://www.esl-lab.com/). Evaluating these sites is more straightforward because a more specific and focused set of evaluation criteria may be appropriately applied.

The Designer-Evaluator: Selected Points of Focus

Many CALL materials are created by language teachers themselves, either working on their own or in close association with those who have the appropriate design or technical expertise, such as multimedia designers or programmers. In such cases, it is very common for the language teacher to be closely involved in the evaluation as well as the design of the materials, and a number of examples are given in this section. The designer-evaluator perspective is a very important one in contemporary CALL evaluation, and it is introduced in this section through representative examples.

The four examples that follow include a description of the project, the approach taken in the evaluation, and the chief criteria on which the evaluation was based. The examples include Web sites and software applications that each have a particular focus when it comes to evaluation. Evaluations have been selected to illustrate a range of approaches and criteria. The studies chosen focused on a language-teaching methodology for culture learning, online teaching with a new CMC technology, a language and a language skill-related project for *kanji* learning, and a group of students and courseware with a vocational orientation. In their different ways, all of these designers were trying to make decisions on how well their CALL materials (methodologies, strategies, technologies, programs, courses) were working.

A Methodology Focus.   Furstenberg, Levet, English, and Maillet (2001) described and evaluated a Web-based, cross-cultural, curricular initiative entitled *Cultura*, which, according to the authors, was "designed to develop foreign language students' understanding of foreign cultural attitudes, concepts, beliefs, and ways of interacting and looking at the world" (p. 55). Specifically, *Cultura* illustrates how the Web can be used to foster cross-cultural understanding between American and French students. In this section, the basics of the design of the project are introduced briefly, and then the focus is on the evaluation.

Culture utilizes a sophisticated, pedagogical approach to learning about language and culture using computer-mediated communication, and a variety of comparative techniques are invoked. These include, for example, juxtaposing video segments in which the same word is used by different people in different contexts in the culture of the L2. The language learner can then reflect on how that word is used differently within the culture of his or her own L1, before going on and making a comparison with the L2. Differences in the use, contextual meaning, and application of key words may then be discussed between native speakers and language learners through an e-mail discussion forum to further expand and increase cultural understanding. The pedagogy involves three main elements: The students complete a series of questionnaires in their L1 on the Web; they engage in individual and group discussions on cultural similarities and differences, at first in their L1 and then through dialogue with partners; and they access a wide range of authentic materials that are used to generate discussion and comment. Through such techniques, *Cultura* has been designed so that students gradually construct an understanding of important aspects of their partners' culture.

The project was evaluated by Furstenberg et al. (2001) using the following criteria:

- Usefulness and interest for cultural understanding;
- Quality of materials and activities;
- Web interface;
- Nature and frequency of resources used; and
- General assessment concerning games in understanding the target culture. (p. 87)

Using a Likert 5-point scale, the students evaluated the usefulness of the questionnaires that had been completed during the project, together with the ancillary texts. The questionnaires played an important role in the design methodology of *Cultura* because they were judged to help students make visible their cultural beliefs and understandings. As a result, an evaluation of the effectiveness of the questionnaires was critical for the designers. Naturally, then, we would expect this methodological aspect of the project to be the essential focus in the evaluation, and such was the case. The questionnaires received the students' highest approval rating, followed by the texts and then the films, so this aspect of the methodology was considered to have worked successfully. Also, incidentally, the discussion forum as a source of cultural information rated very highly. Other aspects of the design of the Web site that the developers were interested in were evaluated as well.

Through the evaluation process, it was clear that the authors' main goal was to evaluate user perceptions and obtain feedback on aspects of the methodology being introduced, rather than to evaluate learning outcomes or to gather comments on the technical design quality. In other words, the evaluation was narrowly defined and decision driven. If assessing the learning outcomes had been the focus, the idea of culture learning would need to have been made more concrete so that measurement of specific abilities or skills could have occurred through 'before' and 'after' testing. Conventional tests or, perhaps more imaginatively, culture games could then have been devised so that a culture learning score could have been generated. In the final section of the evaluation, students were asked to make suggestions for the future. The authors reported that learners requested particularly an increase in the CMC channels available, to include more videoconferencing sessions and a chat facility for synchronous discussion.

An Online Teaching and Technology Focus. *Lyceum*, the virtual learning environment (VLE) mentioned in chapter 2, has also undergone extensive evaluation (Hampel, 2003). The particular nature of the audiographic conferencing tool (especially the particular combination of communication modes) and the purpose to which it was used in online tuition both contributed to the choice of evaluation criteria and process. As described earlier, as far as the tool itself is concerned, it has quite a large number of components, including the whiteboard facility and text chat. As a result, an evaluator necessarily needs to know how well each of these components is working during online tuition, both separately and in combination. Such tools are very new, not only to the learners, but also to the teachers or tutors, and thus their attitudes and responses to the tool as a pedagogical or learning instrument need to be incorporated into the evaluation process as well. Furthermore, the evaluation needs to include some measure of how difficult it is to learn how to use the tool, and how much time is required before inexperienced users can feel comfortable and confident. These are just some of the questions that designers or users of new tools might seek to answer through an evaluation.

In this example, the stage the tool had reached in its development was a consideration as well. Hampel (2003) reported that, in 1999 and 2001, *Lyceum* was trial tested for use in French and German language tutorials, before being introduced in early 2002 in the mainstream German course. Again, for the evaluator, there are questions that might have been necessary to ask at this particular stage, that would not have been appropriate earlier in the development, and, similarly, might not have been necessary to ask.

In the evaluation described here, which was devised in early 2002, Hampel's (2003) main goal was to evaluate tuition via *Lyceum*. Specifically, she focused on the "viability" of online tuition and the "experience" of students and tutors. She accomplished this through observing one of the tutorial groups (February to July) and through the assistance of student and tutor volunteers who kept logbooks reflecting on their online tutorial experience and who answered a questionnaire toward the end of the observation period. The findings of the evaluation were reported from the student and tutor perspectives in terms of benefits and drawbacks. Benefits included a perception of improvement in the students' aural communication skills and the increased opportunity for language practice. Drawbacks included technical difficulties and responses to them, the lack of body language in the VLE, and the demands of the interface and other contextual issues.

Interestingly, the multimodal nature of audioconferencing, reflected in the range of tools available, was perceived to be a benefit and a drawback: On the one hand, different tools were found to be used for different purposes; on the other hand, some students said there were too many things to think about at the same time. This is not unusual with newcomers to any new technology, even much simpler ones, such as e-mail. What is important in the context of online tuition and *Lyceum* is that specific improvements to the system have been made on the basis of the student and tutor feedback. For example, a mechanism has been built into the system to make turn taking easier, special emoticons have been introduced, and the availability of technical help has been improved.

In this example, then, we see that an integrated form of evaluation took place. Although technical considerations were clearly a focus, they were only a focus in the context of online tuition. In other words, the interdependency of the pedagogy and the technology was recognized, and the functionality of the system from the technology perspective only mattered inasmuch as it impacted on the tutor and the student in online teaching and learning.

It is worth noting that the work of Hampel (2003) follows a sustained effort at the Open University in the evaluation and development of online learning tools. While *Lyceum* was still in the design and development stage, a group at the university had undertaken a very detailed evaluation of online tools for distance language learning (see Hewer, Kötter, Rodine, & Shield, 1999). Hewer et al. (1999) developed specific criteria for evaluating online fluency-oriented tasks using synchronous and asynchronous CMC tools employing a range of functions (e.g., audioconferencing, whiteboard, chat). They looked at learning and technology considerations plus "practicalities" such as cost, licensing, availability, intuitiveness of use, documentation/support available, and the technical requirements needed to run the software. Thus, when *Lyceum* did eventually become available for comment and use, staff associated with the project had built up a considerable base of knowledge and experience that could be used to inform the project.

A Language/Language Skill Focus. In chapter 2, we saw that CALL designers sometimes used frameworks specific to a language skill or language area, and that sometimes characteristics of the language itself became important. This is also a feature of CALL evaluations. There are many instances in which a more narrowly defined framework derived from theory or best practice in the domain of the language skill or area is better suited for evaluating new CALL materials than is a general-purpose framework. For example, Komori and Zimmerman (2001) focused on evaluating five Web-based *kanji* programs with specific reference to features that promote autonomous learning. Their criteria were developed from the literature on autonomous *kanji* learning and on a critical review of the effectiveness of *kanji* learning programs.

The authors stated right at the outset that it is the problem of learning so many *kanji* in the development of reading proficiency that motivated their project. On the basis that *kanji* learning is similar to vocabulary learning for alphabetic languages, it is this literature that formed a central reference as Komori and Zimmerman developed their evaluation criteria. This material was used alongside the work completed on autonomous learning strategies. Having formulated their approach, they set out to examine five *kanji* learning programs on the Internet for "acquisition-promoting features related to phonology, orthography, practising, grouping, memorising, referencing and contextualisation" (p. 51). In this, they paid special attention to the phonological features, such as the *on* and *kun* readings and sound files of the pronunciation, and the orthography, including the number of strokes and the stroke order. All these qualities and characteristics come into the foreground in online *kanji* learning. In the discussion of the strengths and limitations of the programs, as well as responding to the stated criteria, they also picked out commonalities and differences, and commented on particular features of individual programs; for example, the drill and practice function of the *Java Kanji Flashcard 500* program, or the *QuickTime* animation movies showing stroke order in "Gahoh." Similarly, drawbacks or problems with individual programs were highlighted. The evaluation was presented through prose rather than a checklist picking out key points.

Komori and Zimmerman (2001) drew the discussion together by deriving from their evaluation of the five *kanji* learning programs, features that they wished to add to their own program, *WWKanji*. In this way, their systematic approach to the evaluation of a group of similar programs helped shape and improve the design of their own program.

With Blok, Van Daalen-Kapteijns, Otter, and Overmaat (2001), a framework for the description and evaluation of word learning courseware was formulated, and again five programs were evaluated. The authors developed a set of six relevant characteristics and factors to describe desirable teaching qualities of courseware for word learning. The first three characteristics and factors were word-learning specific, whereas the second three factors related to language learning more broadly. The authors then developed evaluative questions for each of the six relevant characteristics. For example, the three questions for the category "learning goals" were:

- Which criteria are used for the selection of the target words?
- How many words does the program treat?
- Are there any explicit goals regarding the pursued knowledge level (including receptive or productive use)? (p. 111)

Blok et al. argued in their conclusions and discussions that the framework was successful even though the five programs were different from one another in important ways, for example, some focused on incidental vocabulary learning, whereas others pursued intentional word learning. Blok et al. noted that it was only the feedback and evaluation criteria that returned a poor result in this case—it did not differentiate among the programs sufficiently—because the programs concerned did not offer facilities for feedback or self-testing.

Other pertinent evaluation articles with a skill or language area focus include Wood (2001), who evaluated 16 vocabulary programs for children, and Nesselhauf and Tschichold (2002), who also concentrated on vocabulary and seven commercial programs. In contrast, Neri et al. (2002) provided a comprehensive evaluation framework for pronunciation systems.

A Student/Courseware Focus. In CALL, often an evaluation arises out of the need to assess the effectiveness of a course or courseware that has been developed. Such was the case in the *Airline Talk Project*, which was part of the European Leonardo da Vinci project, devoted to vocational training (Gimeno-Sanz, 2002). The goal of the project was to provide language-learning materials—in English, German, and Spanish—that could promote continuing, autonomous learning for airline staff.

What was apparent very early in the design of this project was the effort with which the author and her colleagues set out to understand the needs and characteristics of the students who would use the courseware. A thorough needs analysis was conducted at the outset and used to gather the relevant information for the design of the software. The focus was very much on helping the learners to develop their ability to function in the foreign language rather than to train them in their particular field. Also, emphasis was placed on flexibility, motivation, and learner centeredness. The design was developed around a number of situations or scenarios, ordered chronologically, and linked using the metaphor of the journey, from check in at departure to arrival. Activities included grammar, language functions, context-oriented activities, vocabulary practice, speaking practice, and pronunciation. Multimedia features, such as video sequences and subtitling, were used to introduce the language in the context of a particular scenario.

In this case, a summative evaluation was carried out, and then the results were used formatively to inform a second iteration of the airline talk project, *Airline Talk 2*. The goal of the evaluation was:

> to investigate whether the CD-ROM met the learning needs of the target group, i.e. learners of Spanish employed by the airlines, in terms of content and design; to investigate the reactions of language teachers who had not been involved in the writing of the materials; and to identify any remaining technical or other snags. (Gimeno-Sanz, 2002, p. 53)

The *Airline Talk* project employed formative and summative evaluation techniques. The distinction is an important one. According to Boyle (1997), the former is conducted during the development of the system, whereas the latter is conducted on completion or at a significant staging post in the long-term development cycle, such as in the case here, between Versions 1 and 2 of the project.

The value of formative evaluation is now widely recognized. In the early days of educational technology, often only summative evaluation studies were conducted. This approach meant that users only tried the software and gave feedback when projects were complete, and often little could be done to reconfigure the materials if they were found to be inadequate in some way. In contrast, formative evaluation is "interleaved" through the development process and it is especially important because of the ways in which the data collected as part of the evaluation can inform the actual construction of the program (Boyle, 1997). Formative evaluation techniques may be used to address and test specific design features within the program. As Boyle (1997) observed, "the prototype becomes an object of communication between the designer and the users." (p. 186). This kind of information and feedback from prospective users early on in the development of materials is extremely valuable.

A detailed understanding of the specific vocational needs of the student audience permeates the *Airline Talk* project, both in the design and in the approach taken to evaluation. The second iteration of the project became a closer approximation and response to the specific needs of the students.

This example concludes this section on the designer-evaluator. It is clearly evident that the designer-evaluators' approach is a very particular one, because they know exactly what they are looking for, what aspects of the design should constitute the focus, what decisions need to be made, and what specific information will inform further iterations of the design.

Also, typically, they have a deeper understanding of the learning context and learner needs, and this knowledge further shapes and directs the focus, the criteria, and ultimately the questions to be answered through the evaluation.

Larger-Scale Frameworks

We now move away from the more narrowly focused approaches to evaluation, such as those used by the designer-evaluator, to more generic, broadly conceived frameworks. A number of such frameworks are available in the wider literature on the use and evaluation of IT or multimedia in education (e.g., Barker & King, 1993; Gibbs, Graves, & Bernas, 2001). These general frameworks have not been developed for CALL specifically, and, for a number of reasons, a framework specifically designed for language learning is beneficial (see Hubbard, 1988). Hubbard's methodological framework, which is described here, and Chapelle's framework, which is described later in the chapter, are the best-known frameworks available in CALL that relate to evaluation (Chapelle, 2001; Hubbard, 1987, 1988, 1992, 1996; see also Murray & Barnes, 1998).

Hubbard's Methodological Framework. Hubbard's early work (1987) looked at the role and value of checklists. He emphasized the importance of developing evaluation checklists that were made specifically for second language learning rather than for other subject areas. He also included a critique of the checklists of the time, and stressed the critical importance of the "need for the evaluator to have a clear understanding of the *approach* underlying the curriculum and the syllabus" (p. 230) and the "fit of the software to instructional approach" (p. 229), a point that is often overlooked in general purpose checklists. Hubbard then provided a detailed rationale, an evaluation form for language teaching and learning, and a procedure describing the way it should be used. His evaluation form was in three parts: an approach checklist, a learner strategy checklist, and other pedagogical considerations. The two checklist components were presented in the form of statements, with a space for an evaluator response using a Likert-like scale as a measure of degree, plus space for more open-ended comments. Of the procedure or process, Hubbard (1987) noted:

> The actual evaluation process involves five steps, moving from a cursory level to a very detailed and critical review. As the goal is not necessarily to complete the form but to decide on whether to adopt a given piece of software, the evaluation procedure should normally stop at any point where the evaluator becomes convinced that the software is *not* appropriate for his or her class or program. (p. 249)

This approach was very much decision driven; essentially, the process stopped once sufficient information has been gathered by the evaluator to reach a decision.

Building on this work, Hubbard (1988) went on to develop and present an integrated methodological framework for CALL courseware evaluation. The framework was further refined in Hubbard (1992), which focused chiefly on development, and Hubbard (1996), which incorporated implementation as well. Ultimately, the framework combined three interrelated modules covering development, evaluation, and implementation; the evaluation module is the focus here. Hubbard (1992, 1996) summarized the principles underpinning his framework, and asserted that it should:

1. Be based on or be consistent with existing frameworks for language-teaching methodology.
2. Be "nondogmatic and flexible," and should "not be tied to any single conception of the nature of language, language teaching, or language learning" (1992, p. 42).
3. Link development, evaluation, and implementation explicitly.
4. Identify the elements of the teaching/learning process and the multiple, interrelationships among them.

His framework drew on previous frameworks by Phillips (1985b) for CALL, and Richards and Rodgers (1986) in their description of a language-teaching method. Together, these prior frameworks provided the organizational structure of Hubbard's refined framework (1992, 1996) and were responsible for many of the elements defined within it. The overall structure came from Richards and Rodgers (1986). This model was described at the levels of approach, design, and procedure, and it was nonhierarchical. Briefly, *approach* is the more abstract level and represents a theoretical account, or assumptions, on the nature of language and language learning; *design* represents the realization of the theory or the assumptions in terms of the syllabus or curricula goals, learning tasks, and activities, and the roles of learner, teacher and materials; and *procedure* represents the implementation of those goals through specific techniques and practices in the classroom (Hubbard, 1992; Richards & Rodgers, 1986). When reconceptualized for CALL, procedure embodies a description of the characteristics of a CALL software program, including the answer evaluation and feedback mechanisms.

The main elements in the Hubbard framework, though not their interrelationships, are shown in Table 3.3 at the three levels. One will note the change of names in the title line of the table. For the evaluation module, Hubbard (1996, p. 27) chose to replace the terms used by Richards and Rodgers—*approach, design*, and *procedure*—with the parallel terms *teacher fit,*

TABLE 3.3

Main Elements in Hubbard's (1996) Evaluation Framework

Teacher Fit (Approach)	Learner Fit (Design)	Operational Description (Procedure)
Linguistic assumptions	Learner profiles	Control options
Learning assumptions	Syllabus	Input judging
Language-teaching approach	Language difficulty	Screen Layout
Computer delivery system	Program difficulty	Presentation scheme
Approach-based design criteria	Content	Activity type
	Learning style	Feedback
	Program focus	Help options
	Classroom management	
	Learner focus	
	Hardware and programming language considerations	

learner fit, and *operational description*, respectively. These new labels were felt to give evaluators a better idea of what the category or item encompassed. Once the evaluator had analyzed learner fit and teacher fit for a particular piece of software or courseware, Hubbard saw this information feeding into two distinct areas, which he called "appropriateness judgments" and "implementation schemes" (Hubbard, 1996, p. 28). The first looked at the CALL materials against the backdrop of a particular learning situation and group of students; the second looked at when and how the material would be used with students.

Hubbard's work in this area has been of lasting value (see Hubbard, 1987, 1988, 1992, 1996). It has been referenced frequently (e.g., Burston, 2003b; Levy, 1997), and it forms the basis of a principled and consistent approach to software evaluation that is used in the *CALICO Journal*, which has regularly published software reviews since its inception (see Burston, 2003b; Hubbard, 1987, 1988).

Chapelle's Theory-Based, Task-Oriented Framework.   Like Hubbard's framework before it, Chapelle's (2001) general evaluation framework has been important because of its comprehensiveness and principled approach. The framework has also been used or referenced by a number of CALL authors (Blin, 2004; Hincks, 2003; Jamieson, Chapelle, & Preiss, 2004, 2005). The background theory, principles, and criteria used in Chapelle's framework are covered here in some detail to pave the way for discussion later, especially in contrasting the approaches of Hubbard and Chapelle. The comparison is valuable because the Hubbard framework is essentially methodology driven, whereas Chapelle's framework is theory driven. In fact, this framework is not only theory driven, in the sense that it argues for theory to be the point of departure when creating an evaluation framework; it is also theory specific, in that it is driven by a *particular* theoretical account, largely represented by the interactionist position (see Harrington & Levy, 2001). In contrast, in his evaluation framework, Hubbard did not adopt a particular viewpoint theoretically. The difference in perspective is useful for anyone contemplating a CALL evaluation study motivated by general principles drawn from language teaching and learning.

SLA theory is absolutely central in Chapelle's (2001) evaluation framework, specifically theories related to "ideal cognitive and social affective conditions for instructed SLA" (p. 45). There is a special emphasis on the design and structure of the language-learning task as a means by which theory is put into practice. In this regard, Skehan's (1998) guidelines for implementing effective task-based instruction are invoked, together with research findings drawn from the focus-on-form literature, especially that relating to attentional manipulation, and the belief that directing learners' attention to linguistic form during meaning-focused tasks is the most likely condition to lead to learner acquisition of target language structures (see Chapelle, 2001; Doughty & Williams, 1998; see also Levy & Kennedy, 2004). In this orientation, learning to manipulate the forms of the language is regarded as the core requirement of the language learner. In addition, Chapelle included some further conditions that should be created for successful language acquisition, although the detail is not discussed.

Chapelle (2001) argued that to improve CALL evaluation we must improve the evaluation criteria in three ways: first, by incorporating the most recent findings and theory from SLA on the ideal conditions for language learning; second, by providing guidance on how the criteria should be used; and third, by ensuring that the criteria and the theory apply both to the software and to the task completed. In moving to meet these requirements, Chapelle (2001) presented a perspective on CALL evaluation that embodies five principles, summarized as follows:

1. Evaluation of CALL is a situation-specific argument.
2. CALL should be evaluated through two perspectives: judgemental analysis of the software and planned tasks, and empirical analysis of learners' performance.

3. Criteria for CALL task quality should come from theory and research on instructed SLA.
4. Criteria should be applied in view of the purpose of the task
5. Language learning potential should be the central criterion in evaluation of CALL. (p. 52)

In addressing the first principle, Chapelle (2001) wrote specifically about evaluating CALL tasks and argues that the outcome "cannot be a categorical decision about effectiveness" (p. 53). She continued:

> Instead, an evaluation has to result in an argument indicating in what ways a particular CALL task is appropriate for particulars learners at a given time. In other words, CALL task appropriateness needs to be evaluated on the basis of evidence and rationales pertaining to task use in a particular setting. The idea of evaluation as a context-specific argument rather than a categorical judgement, of course, makes evaluation a complex issue, which needs to be addressed by all CALL users. (p. 52)

The statement is consistent with the approach taken to software reviews in the *CALICO Journal*. Reviewers do not necessarily evaluate CALL tasks specifically, but the evaluation is still presented as an argument. The discussion does not lead to an absolute decision on the effectiveness of the software, but instead creates a sense of its strengths and limitations in a particular context.

With regard to the second principle, Chapelle (2001) outlined two levels of *judgmental* analysis, labeled *CALL software* and *teacher-planned CALL activities*, and one level of *empirical* analysis, called *learner's performance during CALL activities*. Only the first level, CALL software, is decontextualized, and this is also the only level where Chapelle referred to evaluation checklists. The second level concerns the way in which the teacher plans and organizes the use of the software within a language class, and so it is context specific. Similarly, to accomplish empirical analysis at the third level, an appreciation of context-specific factors is crucial; this would include collecting and analyzing data as the students use the software, and, for Chapelle, would involve seeking evidence of negotiation of meaning through interactional modifications. Chapelle says that all three levels of analysis complement one another, and that evaluation conducted at each level contributes to building an evaluation argument.

From this discussion, Chapelle (2001) derived her six criteria for CALL task appropriateness, as shown in Table 3.4. These six criteria for CALL task appropriateness are used for both judgmental and empirical purposes, to generate questions for judgmental analysis of CALL appropriateness, and also to generate questions for the empirical evaluation of CALL tasks.

For Chapelle (2001), *language-learning potential* was the most critical criterion in CALL evaluation; it refers to the extent to which the task promotes

TABLE 3.4

Criteria for CALL Task Appropriateness (Chapelle, 2001)

Language learning potential	The degree of opportunity present for beneficial focus on form.
Learner fit	The amount of opportunity for engagement with language under appropriate conditions given learner characteristics.
Meaning focus	The extent to which the learner's attention is directed toward the meaning of the language.
Authenticity	The degree of correspondence between the CALL activity and target-language activities of interest to learners out of the classroom.
Positive impact	The positive effects of the CALL activity on those who participate in it.
Practicality	The adequacy of resources to support the use of the CALL activity.

Reprinted with the permission of Cambridge University Press.

beneficial focus on form. Task characteristics identified as relevant for promoting focus on form are interactional modifications, modification of output, time pressure, modality, support, surprise, control, and stakes (the learners' perception of the importance of accurate performance). Tasks that do not promote beneficial focus on form have low language-learning potential (LLP), including CALL activities that simply provide opportunities for language use.

The construct LLP also excludes other incidental, noncore areas of language learning, although Chapelle (2001) acknowledged that language-learning tasks engage students in more than the learning of the grammar or the mechanical aspects of the language. She maintains that tasks should also promote or encourage engagement with, or development of, learning strategies, learners' social identity, cultural awareness, pragmatic abilities, and computer literacy. Chapelle grouped these noncore aspects of the task under *positive impact*. This decision concerning what is central to language learning and what is not is fundamental to the character of Chapelle's framework and rationale.

The remaining elements in Chapelle's (2001) schema, their labels and their content, are consistent with established practice, in CALL and more generally in language teaching and learning. Learner fit is the category that is used to account for individual learner differences in all its aspects, both linguistic and nonlinguistic. As such, it is very similar to Burston's (2003a) definition of learner fit, mentioned earlier. For Chapelle, this quality encompassed matters relating to the importance of choosing tasks that enable individual learners to work with language structures at the appropriate level, as well as learner characteristics such as motivation, age, gender, and learning style. The *meaning focus*, now well accepted, requires that the task is designed such that the learner focuses on meaning over form.[1] *Authenticity* relates to the connection between in-class and out-of-class activity, and refers to the degree of correspondence between the two, as far as the learners' interests and needs are concerned. Finally, *practicality* refers to the particular context within which teachers and learners are working. It includes the availability of hardware and software technical support and the social and cultural characteristics of the institutional environment. The institutional infrastructure and many of the integration issues, therefore, as described in the last chapter, would fall into this category.

DISCUSSION

Introduction

In the descriptive section of this chapter, a number of characteristics concerning the products and processes of evaluation studies in CALL came to light. In terms of *what* has been evaluated, we have seen great diversity. The objects of the evaluations and descriptions have included a multimedia program on CD for teaching pronunciation; a Web site for learning English; a Web-based, cross-cultural curricular initiative; an online teaching tool; two vocabulary learning programs; and courseware specifically designed for a particular group of learners. This list might imply that technology choice and design is always necessarily a central feature in evaluations. However, this would be misleading. The focus of CALL evaluations, as we have seen in

---

[1]A focus on meaning is central to the definition of language-learning potential as it is defined here and in the SLA literature. Therefore, the need for a separate category called *meaning focus* may not be necessary, or it may have been included for emphasis. In the recent SLA literature, meaning or communication is always regarded as primary in definitions of focus on form. In the words of Doughty and Williams (1998), "[F]ocus on form entails a prerequisite engagement in meaning before attention to linguistic features" (p. 3). In addition, when a focus on form does occur, typically this arises in a way that is incidental or unplanned. Long and Robinson (1998) wrote of "an occasional shift of attention to linguistic code features—by the teacher and/or one or more students—triggered by perceived problems with comprehension or production" (p. 23). Thus, we are seeking tasks that are primarily communicative, meaning-focused activities, with a provision for an occasional or incidental focus on linguistic form that is initiated by either teacher or student. The problem may arise because of the way focus on form is defined in the SLA literature, in which the nomenclature implies a focus on form whereas the real focus, the primary one, is on meaning, with only an occasional shift to form. Chapelle may also have included the meaning focus category for this very reason, to emphasize this focus given the way the label can easily obscure its meaning.

the examples, is equally concerned with pedagogy, methodology, and effective learning strategies. Also, in terms of *how* these CALL artifacts have been evaluated, we have seen a range of approaches and frameworks adopted, from specialized designer-oriented evaluation frameworks to larger-scale, general-purpose frameworks. In this section, we analyze and discuss the strengths and limitations of these various frameworks and approaches in relation to the technological and pedagogical contexts in which they are applied.

Checklists

Perhaps not surprisingly, because of their simplicity and because of the way they may be perceived to lead the evaluator to address only a series of simple binary decisions, the use of checklists in CALL evaluation has come under considerable scrutiny over the years (see Burston, 2003b; Hubbard, 1987; Susser, 2001). However, Susser (2001) offered a spirited defense of checklists, firstly by identifying the main areas of criticism and then by responding to them. We believe that his analysis has many strengths. There is considerable depth in his argument, so the original article should be consulted for those who are seeking detail. Here, briefly, the main points are presented, after a definition of a checklist given by Susser (2001) as follows:

> I define 'checklist' in terms of format: it normally consists of a series of questions or statements to be checked off 'yes/no' or marked 1–5 on a Likert scale, or has blanks to be filled in. A checklist may be in questionnaire format or accompanied by lengthy text explanations; a series of questions in paragraph form also qualifies, as does a bare list of features to be looked for. (p. 262)

Susser grouped the objections to checklists into six basic categories. These objections and an abbreviated summary of Susser's responses to them (given in italics) are:

1. Problems of accuracy, compatibility, and transferability (Decoo, 1984) caused particularly by demanding overly simplistic categorization; for example, 2-way (yes/no) and 5-way (e.g., Likert scale). *There is no rule saying that the use of checklists precludes provision for free commentary, and in fact many evaluation forms have contained a basic checklist plus sections for a more open ended response as well (e.g., Hubbard, 1987).*

2. A focus on the technology rather than on the teaching and learning aspects. *Previous checklists can and have dealt with teaching and learning issues, as well as technology; in fact, most checklists for CALL have done this (e.g., Levy & Farrugia, 1988).*

3. Lack of objectivity, reliability, and validity. *Susser argued that many of the studies cited relating to this category have suffered from methodological flaws. Some experimental evidence is also favorable to checklists.*

4. An implicit or explicit bias toward a particular approach or method. *True, but this is inevitable for all frameworks. In fact, one could argue that this feature is to be desired because it indicates a consistent and principled basis for the checklist.*

5. The assumption that it is possible to design a general-purpose evaluation tool that can be applied across the board to any software/courseware. *Susser made a threefold response: First, checklists can comprise two parts, one general, one specific; second, if desired, checklists can be designed for particular kinds of software/courseware; third, checklists do need updating regularly as new technologies and new CALL materials become available.*

6. The need for background knowledge and experience to enable accurate and appropriate responses through the process of completing the checklist. *This is always true; CALL evaluators need to develop their skills as in any area of CALL (e.g., design or teaching).*

Susser (2001) concluded that checklists are a valuable tool that can help remind teachers of the wide range of elements that need to be covered in evaluating software and courseware, especially given that there are multiple paths through the materials, some of which are not immediately apparent, unlike a textbook for example. He also observed, following Knowles (1992), that checklists can play an educational role for language teachers in helping them make more visible and explicit those assumptions and beliefs that underpin their CALL practice (see chap. 7). Still, checklists are only a beginning. Although a checklist may be able to assist with the initial selection of software or courseware, teachers will still need to monitor its use closely in class (Susser, 2001) reflecting Chapelle's concern for empirical evaluation as well as judgmental evaluation.

Surveys

We saw earlier in the section on the use of surveys that valuable information could be gathered in this way to assist in the refinement of project design. It was also noted that although surveys have their uses, they also have their drawbacks if relied on too heavily. As a result, a more composite approach to evaluation is preferred by many more-experienced CALL practitioners. Evaluations may certainly include surveys of some kind—in fact, they usually do—but the evaluator does not rely solely on this source of information. An approach that embraces a number of complementary data collection instruments is generally favored because it may go some way to countering the well-known deficiencies of surveys when used in isolation. Hémard and Cushion (2001) provided an informed view in this respect:

> Although, from experience, questionnaires handed out to students as part of the internal monitoring system rarely seem to be effective and useful, it was nonetheless felt that this evaluation method could be appropriately used in conjunction with others to provide further data on students' level of ICT competence, their views on the CALL interface and how they accessed it, if at all, within their own learning context. (p. 20)

Furthermore, Hémard and Cushion's evaluation of their Web-based CALL project took place over a 12 month period and involved a range of data collection techniques, in addition to the questionnaires. These included peer evaluation and discussion in the formative stages of the design as well as user walkthroughs and workshops in the summative phase of the process. The user-walkthrough method is worthy of further note. It is similar to a verbal or think-aloud protocol and is "primarily an evaluation technique designed to focus on the learnability and usability of a system" (Hémard & Cushion, 2001, p. 21). In this project, five walkthrough sessions of 2 hours each were organized over 5 weeks. The sessions enabled the designers to obtain important feedback on such matters as the interface design, ease of use, clarity of the learning objectives, and, most important, whether the design met learner expectations for a multimedia Web-based program. In this regard, considerable care has to be taken with CALL materials that are learning theory-driven because although they may meet the requirements of the learning theory, they may fail miserably in terms of design by using outmoded formats and visual presentation techniques. In this regard, Boyle (1997) noted that presentation design has its own problem space, is quite different from conceptual design, and requires its own theories and rationale. Different facets of a fully realized CALL program may well draw on a number of complementary theoretical bases (see chap. 5 for a fuller discussion of this topic).

Although surveys conducted for evaluation purposes are usually designed by an individual designer or teacher, or for a small, local audience, it is still well worth reporting the results of such evaluations in venues that provide access to a wider audience (e.g., via an Internet discussion list). The power of evaluation studies using surveys (questionnaire/interview), often in combination with other sources, lies in the parallels that may be drawn by readers who are following the same line of thinking or pattern of development work. There is much to be gained by reading of another's experience, especially if reported perceptively, generously, and in sufficient detail. For example, for those developing online courses (as many are in CALL these days), evaluation studies highlighting specific design problems and solutions, student attitudes and perceptions (especially when given differentially according to, say, ability level), motivational effects, learning preferences and so on can be extremely useful for others. The value to the reader of reports of evaluation studies such as these is extracted in a rather different way from the traditional research study report that aims at generalizability through a rigorous research design and precautions taken to ensure the validity and reliability of the results. In evaluation reports of the kind described here, the power of the report derives from the potential for shared experience. The best contributions convey an in-depth understanding of the problems that designers face, and offer wise and informed ideas for their solution.

Designer-Oriented and Third-Party Evaluations: Strengths and Limitations

Typically, the designer-evaluator has an intimate knowledge of the CALL materials involved in the evaluation, the nature and characteristics of the anticipated student audience, and the learning environment or context in which the materials will be used. Such was the case in the four designer-evaluator examples given earlier in the chapter.

By contrast, the third-party evaluators, language teachers, or software reviewers are working in the dark. They haven't been closely involved in the design and development of the CALL materials, so they have to spend considerable time getting to know them first. Also, software reviewers are unlikely to have a detailed knowledge of the audience; instead, they rely on their knowledge and experience of CALL, and language learning more generally, to make judgments on how successful the program might be when actually used with students.

The designer-evaluator studies described earlier were conducted "in house" by the designers of the projects themselves, or people closely associated with them. The designers, by allowing the rationales for the projects to determine their goals, knew exactly what questions they wanted to ask when it came to the evaluation. This meant that the criteria for each project could be precisely defined in order to answer the questions in which the designers were primarily interested. In this regard, Furstenberg et al. (2001) provided a good example. The feedback through the evaluation was sufficiently focused and directed to enable the authors to make specific refinements to the approach and the resources available. The authors also went beyond evaluation per se by pointing toward assessment-oriented research questions that must be answered to identify the most significant elements that help students construct true cultural understanding; thus, an initial evaluation study is helping to shape the more in-depth research agenda that follows. What emerges in these projects—something that is very significant for evaluation—is the clarity and precision of the goals of the evaluation. This is consistently combined with an understanding of the needs and characteristics of the students and the learning context, including the specific qualities, strengths and limitations of the particular technologies in use.

When third parties evaluate CALL materials, they are at a disadvantage. The language teacher may not know the materials in depth or be able to grasp their potential and, generally, is not able to make adjustments if any aspect is not working as it should. The software reviewer is not able to provide context-specific arguments like the designer-evaluator or language teacher can. The reviewer can only assess the artifact or product out of context, not as a fully integrated component of a course or as a part of regular practice. Instead, the reviewer imagines or projects onto the software how it might be used and what its deficiencies might be when used in actual practice.

Nevertheless, third-party evaluations are still very important. After all, no one can afford to buy all the software that becomes available, or visit and evaluate all the language-learning Web sites, let alone test them in depth with their students. Most language teachers simply do not have the time. Independent, well-reasoned reviews can be very helpful, especially as a first step (following Chapelle's Level 1) in helping language teachers and administrators obtain a picture of the goals, strengths, and drawbacks of the software. Reviews of similar products—both tutors and generic tools used for language-learning purposes—are also most helpful because they assist the potential consumer to narrow the field by comparing programs of similar type, thereby preparing the way for readers to make informed decisions. Also, a word should be said on behalf of software reviewers. They are often very experienced CALL practitioners with an extensive background knowledge of the field. They frequently combine advanced technical expertise with in-depth pedagogical knowledge, so although they may lack an intimate knowledge of the product in use they can still provide valuable expert opinion.

In the earlier discussion of CALL materials like the *Connected Speech* CD or *Dave's ESL Café*, specific information about the learning context was not available. Nonetheless, we still believe the software review, and reflective analysis on characteristics, strengths, and limitations to be most valuable. It is possible to make general statements about the needs of students according to their backgrounds and proficiency levels. It should also not be forgotten that the reader also has knowledge and expertise, and can make informed judgments on the applicability of the information and arguments presented for their own situation.

In much recent CALL activity, the individuals conducting an evaluation are the same individuals who have been involved in the project from the start. Combining the two functions of design and evaluation has both strengths and drawbacks. The strengths lie in the ways in which design has moved to incorporate formative evaluation by adopting a more iterative approach and by incorporating user or learner feedback on design elements much earlier than was formerly the case. Undoubtedly, this has led to better designs that are far more attuned to the goals, needs, and characteristics of learners.

More formally organized third-party evaluations also have their place. Third-party evaluation studies help ensure a systematic and independent evaluation process that is not colored by the predisposition or implicit assumptions of the designer(s). In the CALL literature, there are occasions, in reports on evaluation, when there appears to be a glossing over of the weaknesses in the design or the drawbacks in the approach. If the reader knows that the evaluation was conducted by some independent party, there is perhaps a greater chance of an unbiased assessment of the software or learning environment. Unfortunately—and more experienced CALL designers are very well aware of this fact—repeatedly funds are often not made available for a proper evaluation study to occur. Still, note that some independent evaluations in CALL are reported, as with Söntgens (2001), in which an independent research unit at the institution was commissioned to undertake an evaluation.

The Nature of the Object of the Evaluation

It was noted earlier that the nature of the object of the evaluation was important in choosing suitable criteria. Again, it is worth considering the two examples introduced earlier. Whereas in the first example, the *Connected Speech* CD, the content and form of object of the evaluation were relatively coherent and clear, in the second example, *Dave's ESL Café*, the composite nature of the Web site meant that an evaluation was a more segmented and complex activity. This is by no means the case for all language-learning Web sites—some are very straightforward in their goals, activities, and organization—but Web sites can pose a problem when they include a varied range of activities and interaction types. Each element might require rather different criteria in any appraisal. Although some elements may work very well, others may be seriously defective.

With a commercially produced CD, generally the materials have been designed to form a coherent, self-contained package, usually with a hierarchical structure that can be navigated easily. If the design is strong and the navigational structure clear, the user should be able to view the components of the CD without too many difficulties, although, admittedly, the number of levels and pathways through the material may still pose a challenge. With a program like *Connected Speech*, we have a reasonable sense of the content of the program right from the start—by scanning the title, by reading the small accompanying booklet, or by having a brief look at the CD itself. Also, the content tends to be chosen such that it may readily be conceptualized as a whole with a unifying theme. Thus, when it comes to evaluation, it is easier to obtain a clear sense of what is to be evaluated and to choose the relevant criteria.

In contrast, the boundaries of one of the more complex language-learning Web sites can be difficult to pin down with any precision. For Web sites, the demarcation lines are not unambiguously drawn out for the convenience of evaluators. The people involved can be equally vague: Many authors may contribute in the construction of a Web site and many more may assist in its use, as in *Dave's ESL Café*. Furthermore, the affiliation of the homepage may be clear, but from there the links take the user ever outward to other locations and sites that may or may not be officially part of the original site or recommended by the site authors. Web site names also tend to be a little more open ended, or even metaphorical, like the ESL café, so as not to appear too narrowly defined in an Internet environment that facilitates movement from place to place. For a Web site, then, the content to be evaluated may not be as clear-cut as it is for a CD. Web sites can "hold" many more language-learning materials than can a CD, and they can be diverse in form and function, often including a wide range of tools and resources. When a Web site is evaluated, as a whole, then, evaluation results may appear to be contradictory, as some elements turn out to be more effective than others.

In the *Dave's ESL Café* Web site, some elements are relatively static in their content (e.g., the photos or the phrasal verbs), whereas other elements are dynamic tools that involve student–student or teacher–student interaction, such as Chat Central or the Help Center. These tools are forms of computer-mediated communication (CMC). In a CMC setting where a learner is working with other students or one-to-one with a tutor, clearly the learner's experience will be determined by the quality of the interaction. Results will vary not because of the technology, but rather according to the quality of the interaction and its value for language learning. An effective evaluation of that particular Web site element should be able to reflect the learners' experience. Clearly, relatively stable text-based material needs to be looked at in a different way from elements that support learning interactions. In evaluating a Web site of this kind, there needs to be room for different sets of criteria to be formulated and used for different elements in the Web site.

Web sites like *Dave's ESL Café* are another kind of CALL hybrid, similar in some respects to learning management systems like *BlackBoard* and *WebCT* (see Strambi & Bouvet, 2003). They each involve a number of technologies packaged together. Often, for evaluation purposes, the most effective approach is to treat the parts separately when they offer qualitatively different levels of interactivity, and it should not be considered surprising if some components rate very highly whereas others do not.

CALL materials such as the ESL Café Web site are not single entities, although they may share a collective title. The fact that the elements are gathered together under the same roof, so to speak, can tend to disguise the fact that these significant differences exist. The evaluator needs to ensure that each Web site element is evaluated against a set of criteria relevant to that category. Some elements may be difficult to assess judgmentally (Level 1), because their qualities only become apparent through their use (e.g., chat, discussion forums). They do not contain language content as they stand. This brings to mind a comment by Hubbard (1987) who, in the development of his evaluation framework, noted the distinction between evaluations directed at materials that include language content versus those that are aimed at evaluating computer tools, like the word processor, and the need for rather different approaches. This point was echoed in Levy (1997) in all sections regarding evaluation.

Levy (1997) argued that tutors and tools need to be evaluated against different criteria. Thus, in the ESL Café Web site, the Quizzes section—which is basically a simple computer tutor that evaluates student input—needs to be looked at in a different way from the discussion forums, Chat Central, or the Help Center, which are all examples of computer tools that facilitate student–teacher interaction rather than student–computer interaction. When evaluating computer tutors, the quality of the input processing and computer feedback is crucial, so we would expect these aspects to play a central role in any evaluation. Thus, as noted earlier, Heift (2001) looked at whether students read feedback in a Web-based language tutoring system, and how exactly students work through the error correction process (Heift, 2002; see also Pujolà, 2001, 2002).

With computer tools, Levy (1997) mandated for a two-stage process. First, the tool needs to be evaluated in relation to other computer tools of a similar kind. This is exactly what Hewer et al. (1999) accomplished when they compared synchronous and asynchronous CMC tools prior to the adoption of *Lyceum*. Levy referred to this initial stage as assessing the qualities of the system in relation to other systems of the same generic type. Second, the tool needs to be examined in terms of its context of use for language learning.

When using CMC tools for online learning, it is the nature and quality of the interaction that needs to be at the center of any evaluative assessment. Evaluation of CMC tools was addressed in a paper by Benigno and Trentin (2000) on the evaluation of online courses. They argued for a customized approach given the special characteristics of such courses: "[T]heir quantitative/qualitative evaluation calls for the adoption of specific procedures to assess both the learning process and the participant performance" (p. 259).

Benigno and Trentin's framework for the evaluation of online courses focused on the following aspects, which they labeled "what to evaluate":

- Participants' individual characteristics.
- The participative dimension.
- Message analysis and evaluation from the viewpoint of contents and of collaborative work.
- Analysis of interpersonal communication.
- Effectiveness of the support offered by tutors and experts.
- Participants' reaction to the methodological approach used in running the course.
- Utility of the learning material.
- The learning environment in all its forms—local, virtual, social, etc.
- Communication technology.
- Return on investment compared with similar face-to-face courses.

One can see from this list that it is not so much the language content that is the focus of the evaluation (as in a narrowly defined, language-specific aspect such as "focus on form"), but rather aspects concerning the quality and dimensions of the interactions between participants. This includes levels of participation, the content and quality of the messages exchanged, interpersonal relationships, and the effectiveness of the support offered by human tutors.

General Evaluation Frameworks: A More Detailed Analysis

In any evaluation framework, what will ultimately characterize the framework is the choice of criteria, the relative weighting of the criteria, and the procedure for working through the criteria to reach a result of some kind. The issue of weighting elements is basically avoided in most checklists and it is a recurring criticism (see Burston, 2003b; Susser, 2001). In most checklists there is no indication that any particular section or question is any more or less important than any other. Aside from the natural order, which gives items at the beginning priority over items later on, often there is little else to signify priority order. Here is where the Hubbard and Chapelle frameworks differ substantively from checklists. They each give weight to certain elements over others, and they give order to certain processes. These are the primary ways in which they are more sophisticated evaluation instruments than are checklists. Both of these frameworks also take care to label the elements carefully and provide specific definitions for them, and they supply details relating to the process of evaluation and the range of applicability.

Applying weighting to specific criteria obviously privileges some elements over others. Often, the privilege given to an element or elements directly relates to the philosophy or theory that underpins the evaluation framework, and it is in this territory that differences of opinion can arise. This is not to say that applying a preference to some factors over others is not a good idea; in fact, we believe this is a necessary step. However, in the modern world of language teaching and learning, and with many aspects competing for the teachers' and the learners' attention in language learning, prioritizing certain criteria in an evaluation framework can sometimes be controversial. Criteria that some may wish to promote or bring into the foreground, others may wish to demote to a secondary position, so that they remain in the background; in other cases, the evaluator may want to treat two elements equally. Let us consider Chapelle's (2001) framework as an example, and reflect on the ways in which the priorities are set.

Chapelle's (2001) framework grew from a theoretical orientation that specifies assumptions about what language is and how it is learned. From this theoretical position, six categories of evaluation criteria are derived, with an order of priority. The six criteria are used for judging CALL task appropriateness. The quality called language-learning potential (LLP) is the most important, and it is defined to be the degree of opportunity available for beneficial focus on form. It is the focus on form aspect, and its priority status, that primarily marks the theoretical orientation. This priority is stated explicitly, and it is the foremost decision taken in the construction of the evaluation framework.

There are a number of initial points to note in the development of the evaluation framework. The framework is driven primarily by a particular theoretical orientation known as focus on form (see Doughty & Williams, 1998). This theoretical orientation represents one of a number of alternative second-language learning theories (see Jordan, 2004; Mitchell & Myles, 2004). Chapelle made a choice on what theoretical model to adopt and prioritize in the evaluation framework. It is also noteworthy that although focus on form attracts current interest in SLA research, in many ways it is still controversial, especially with language teachers (see Doughty & Williams, 1998).

The point here is to clearly acknowledge that a series of decisions were made in the development of the framework: first, for it to be theory driven; second, for it to have a single-priority criterion; and third, for it to focus on form during meaning-focused tasks. It is valuable to see how an evaluation framework grows out of a particular theoretical orientation in this way. It is also instructive in the sense that it may guide us in developing other evaluation frameworks that grow out of different theoretical positions, perhaps drawing on two or more theories simultaneously.

Chapelle prioritized focus on form, and held that this is the most likely condition to lead to learner acquisition of the target language structures. In accordance with this theoretical position, learning to manipulate the grammar of language in a meaningful context is regarded as the core goal of language learning. In making this decision, following SLA theory, the priority in the framework is given to learning to manipulate the forms of the language, principally its grammatical forms. As a consequence, other levels of language, pronunciation, vocabulary, or discourse are subordinated, and they are not explicitly mentioned. This point was noted and extended by Neri et al. (2002) with regard to computer-assisted pronunciation teaching (CAPT):

> Although valuable criteria have been outlined in the past few years to evaluate CALL, these are either of a general nature (as in Chapelle, 2001) or they mainly concern computer assisted learning of vocabulary or grammar, while pronunciation is hardly mentioned. This scarcity of indications makes it hard for CAPT practitioners to develop effective courseware. (p. 443)

The LLP also excludes other incidental, noncore areas of language learning, although Chapelle (2001) acknowledged that language-learning tasks teach more than an ability to manipulate the language forms for communicative purposes. She grouped these noncore areas under the heading "Positive Impact" (PI). The way Chapelle distinguished between the principle criterion or "quality" of her evaluation framework, LLP, and the less significant criterion of PI is fundamental to her rationale. It is a way of prioritizing the criteria in the evaluation framework: In this case, one factor is given priority over all others and placed in the primary LLP category. A series of other factors—spin-offs in the process of task completion, one might say—are placed all together in the secondary PI category. This decision and this arrangement largely characterize the evaluation framework.

The priority given to the development of the ability to manipulate the formal aspects of language is by no means universal among language teachers. For instance, in *Cultura* Furstenberg et al. (2001) asserted that they "attempt to redefine the meaning of foreign language 'teaching' in the new world of networked communication" (p. 55), and commented on the rise of the focus on culture in language teaching and learning. They continued:

> Yet, language pedagogy usually still focuses primarily on the mechanics of language skills and devotes little time to the real task of developing students' understanding of another culture, and particularly those aspects of culture that relate to attitudes and values. (p. 57)

> But we owe it to our students to go beyond the mechanics of language and delve, head-on, into the world of cross-cultural literacy. We would be remiss if we did not. (p. 95)

There are many others using CALL who have also emphasized the importance of the development of skills beyond the learning of the language per se. For example, Sengupta (2001) stressed the personal and social dimensions, including learning how communities use language and learning about the world via the language. In addressing students' experiences with new technologies, Warschauer (2000a) also described their value, not principally in terms of their contribution to second-language learning, but as developing important new literacy and life skills that combine technology and language. If one also includes the development of learning strategies, learner autonomy, and so on, it is no surprise that language teachers and learners are having difficulties trying to find the right balance between the many competing elements that authors and commentators argue are important in responding to the needs and goals of the contemporary language learner.

These comments demonstrate that the goals of language teaching and learning are both multiple and changing. Also, views on the priorities for language teaching and learning vary among language teachers and learners, and there are significant differences between the goals of teachers and researchers (see Ellis, 1994; chap. 5, this volume). In tracking these changes and differences, it is helpful to appreciate evolving views of the language-learning task because of its pivotal role in helping researchers and teachers to conceptualize, structure, and motivate language learning. Here, Ribé and Vidal's (1993) description is useful, as noted in chapter 1. They wrote of first-, second-, and third-generation tasks. First-generation tasks were aimed solely at developing the students' communicative ability in a specific area of language. Second-generation tasks developed not only communication skills, but also general cognitive strategies (e.g., analyzing what information was needed in order to complete the task), and included the idea of using language for a 'real' piece of work that had value outside the classroom. Third-generation tasks extend this idea even further. They aim not only to activate communication and cognitive strategies, but also to enrich the students' personal experience more broadly. Third-generation tasks have "a high degree of task authenticity, globality and integration of language and contents and involvement of all the aspects of the individual's personality" (Ribé & Vidal, 1993, p. 3).

Of course, Chapelle (2001) recognized many of these developments, including the broader advantages of CALL use and the students' desire to learn more about the world and to develop their computer literacy—personally, interculturally, and on the computer. Here, the point of note is the way Chapelle weighed these elements in her evaluation framework. They are not allocated individual categories of their own and are not considered primary; instead, they are grouped together in a large category of disparate elements called *positive impact*. In many ways we support Chapelle's decision here, although there may be value in separating out the elements grouped together under PI and priorities may also vary according to one's theoretical orientation, or skill focus. One could also promote learner fit to equal first position to record the fact that responding to individual learners' needs and differences was a primary concern. For Hémard (2003), user acceptability would be prioritized, following HCI principles. Arguably, if learners don't use/accept online language-learning materials, then all else falls by the wayside, including any possibility of language learning through that means. For online CALL, Hémard (2003) stipulated, "successful accessibility must rest on the user's willingness to interact with it [online CALL] and use it as a valuable learning support. On this premise, online CALL activities must not only be useful and meet students' needs but must also be sufficiently enjoyable to be accessed outside the classroom" (p. 40).

In any case, whatever the precise order of priority, it is certainly not as clear as it may have been a decade or so ago that one can confidently extract one element and then treat it as a distinct priority for evaluation purposes. In Chapelle's (2001) framework particularly, the robustness of an SLA theory is called into question. One needs a great deal of confidence in its value and effectiveness in real, as opposed to experimental, learning settings to give it pride of place over a range of other elements that teachers and learners regard as important. Also, in any evaluation study, particular elements will tend to shuffle to the front or move into the background on specific occasions, according to the goals of the study and the specifics of the students and the learning context. However, be that as it may, we believe the great value of Chapelle's evaluation framework is that it provides a valuable template for thinking about these issues, and if the focus in any evaluation study is on CALL tasks with a focus-on-form emphasis, we feel it could be enacted most effectively. It also has the potential to be adjusted or adapted for different purposes, which is undoubtedly a strength as well.

Contrasting the Hubbard and Chapelle Frameworks

The evaluation frameworks developed by Hubbard and Chapelle are easily the most coherent and sophisticated developed so far for CALL. As such, it is informative to contrast their rather different approaches to evaluation. This comparison should not be taken too far, because the frameworks were designed for different purposes, but there is value in setting one beside the other and highlighting some key differences. To obtain a clear sense of how these different priorities give different complexions to the evaluation frameworks associated with them, we need to remind ourselves of the original priorities and definitions.

In Chapelle's (2001) framework, LLP is unequivocally the priority and it is defined as "the degree of opportunity present for beneficial focus on form" (Chapelle, 2001, p. 55). In addition:

Given the importance of focus on language for language acquisition, characteristics among those Skehan identified as relevant for promoting focus on
form—interactional modification, modification of output, time pressure,
modality, support, surprise, control, and stakes—need to be considered in
an argument for language learning potential .... Moreover the complete
meaning of language learning potential will develop as theory and research
in SLA develop .... (p. 55)

In Hubbard's (1996) framework, rather than a single priority, there are
three areas that are considered more or less equally. In the framework,
Hubbard referred to the operational description (procedure) as "the first
step" because it is more objective; then the more subjective judgments of
teacher fit (approach) and learner fit (design) may follow (1996; see also
Richards & Rodgers, 1986). According to Burston (2003b), teacher fit is the
most critical component of software evaluation, and the most difficult to assess. For the purposes of this discussion, however, we treat the three areas of
the evaluation module as being of equal importance, while at the same time
recognizing that one or the other might take priority with a particular
evaluation purpose and setting.

In contrasting the frameworks by Hubbard and Chapelle, it is noticeable
that Hubbard's definitions of teacher and learner fit are more broadly conceived, whereas Chapelle's definition of LLP is narrower, explicitly theory
driven, and directed at evaluating CALL tasks specifically. Although the
definition of teacher fit certainly embodies theory, it does not advocate one
particular theory. It does not give precedence to theory over other elements. It also brings into the equation, at a priority level, classroom methodology and the teachers' curricular objectives. Burston (2003) noted:

> An assessment of teacher fit primarily involves looking at the theoretical un-
> derpinnings of student activities; judging how well they conform to accepted
> theories of cognitive development, second language acquisition, and class-
> room methodology; and determining how closely they accord with the
> teacher's curricula objectives. (p. 38)

Interestingly, both Hubbard and Chapelle shared a similar conception
of learner fit. Given this common ground, it would seem that Hubbard was
prioritizing teaching and methodology, whereas Chapelle was prioritizing
a theory, and, further, a particular theory, as the principal driving force in
her evaluation framework.

Hubbard's evaluation framework is a template that describes interrelationships. This is consistent with Hubbard's second principle for his framework, about which he maintained:

> It should not be tied to any single conception of the nature of language, lan-
> guage teaching, or language learning, nor any specific combination of hard-

ware, nor any specific language skill or mode of presentation. Rather, it
should specify the logical relationships between learners, teachers and computers in as neutral a way as possible. (Hubbard, 1992, p. 42)

In contrast, Chapelle's (2001) framework focuses on establishing priorities rather than identifying relationships. It is also preset at the approach
level, where the theory of language learning is specified explicitly.

In reflecting on the most appropriate uses of the two evaluation frameworks, the Hubbard framework seems to be most suited to evaluating
courseware, or the tutorial components of a Web site (e.g., the Quizzes section in *Dave's ESL Café*), in circumstances in which the program presents
language-learning material, asks questions of it, evaluates learner input,
and provides feedback in some way. Within this framework, the elements
listed under the operational description (procedure) are helpful in pointing to the particular attributes in a piece of software relating to input judging, control, and feedback. These interactional aspects are crucial in the
evaluation of any tutorial CALL activity, and the sophistication with which
they work determines directly the quality of the program or activity.

Looking through the software reviews of the last several years in the
CALICO Journal, this hypothesis seems to be borne out. Most software reviewed is of a tutorial nature, often providing language-learning practice,
in which questions are asked and feedback is given. There are exceptions,
when software reviews of CALL tools are included; for example, reviews of a
translation tool (Cancelo, 2002), a writing support tool (Turnbull, 2002),
and a concordancing tool (Stevens, 2002). The framework remains workable, but the feedback component of the evaluation module is generally not
applicable and new categories are introduced that relate to the specific
qualities of the computer tool. Thus, in the review of *English Spanish Interpreter Deluxe* (Cancelo, 2002), there were significant sections on "Machine
translation systems" and "Evaluation of engine performance." The review
of *Concordance, Version 2.0* (Stevens, 2002) addressed the features of the
concordancer as a research tool. Compared to many of the tutorial-type
programs, the sections on teacher fit and learner fit were relatively short,
and the perspective taken in the review was rather different. There was
more of a focus on the qualities of the tool in relation to other tools of a similar kind, and on describing the kinds of tasks and activities for which the
tool could be most appropriately used with language learners.

In contrast, Chapelle's (2001) framework is directed at the language-
learning task, as it is actually carried out in a particular learning setting.
The framework allows for the analysis of the task at three levels. In tutorial-
type software, this would include: consideration of the task, as presented by
the program per se (Level 1); the particular way the teacher planned and
organized the use of the program in class or out of class, perhaps with pre-

or post-CALL task activities (Level 2); and a careful consideration of what the students actually did when they used the program in context (Level 3). This perspective allows for tasks to be set by the computer program alone, or by the teacher, or a combination of the two. It allows for tutorial- and tool-type programs because the focus is on what students do as they complete the task. However, it would only direct the evaluator to consider input judging, or the feedback mechanisms of a tutorial program in the broader context of meaning-focused task completion.

In tool-type programs, in which the task resides outside the computer program (see Levy, 1997), an evaluation under this scheme would simply revert to an analysis of the task itself; it assumes that the technology tool is transparent and does not exert an effect on the task (see Levy, 2005). It would not require evaluators to compare computer tools of the same generic type (e.g., Hewer et al., 1999). Nor would it lead directly to a consideration of any technology or software characteristic such as those often mentioned in relation to evaluating multimedia CDs or Web sites, including ease of use and navigation; menu structure and design; information accuracy and completeness; use of color, graphics, or animations; media integration, and overall functionality (e.g., Hémard, 2003). To summarize the basic differences between the two frameworks, they are presented in Table 3.5.

CONCLUSION

There is no doubt that evaluation in CALL has become increasingly complex over the last decade. This change has occurred for a number of reasons: a dramatic increase in the technology options available, used separately or in combination (especially with CMC options); a desire to create more complex products, such as Web sites with multiple elements; the increased use of LMSs, which are themselves composite objects; and a wish to undertake more complex and focused evaluation studies embracing the software not only as it stands, but as it is actually used with students. This array of options requires a much more sophisticated approach to evaluation in CALL, especially in choosing the appropriate methodology to suit the specific goals of the evaluation and the technologies in use.

There is a considerable diversity in approach, as we saw both with the evaluation instruments—the checklists, surveys, and general frameworks—and among the designer-evaluators who carefully choose their evaluation goals and approaches to answer very specific questions concerning the use of their CALL materials. Designer–evaluators tend not to use general evaluation frameworks, because the criteria sets and priorities are not designed for the specific needs of their projects. In this sense, evaluation studies are not only context sensitive in terms of the particular students and the learn-

TABLE 3.5

Differences Between Hubbard's and Chapelle's Evaluation Frameworks

	Hubbard (1996) (Burston) (2003b)	Chapelle (2001)
Framework's key features		
Principal purpose	CALL courseware development, evaluation, and implementation	CALL task evaluation and research
Principal object	CALL software/courseware with language content	CALL tasks
Principal orientation or philosophical base	Language-teaching methodology	The interactionist theory of SLA
Principal criteria	Operational description, teacher and learner fit equally	Language-learning potential (focus on form)
The evaluation process		
Product/process emphasis	Product—a specific process not proposed (with a minor reservation—see Hubbard, 1996)	Process—a specific process proposed (three levels of analysis; judgmental and empirical)
Implicit assumptions		
The role of the teacher (typically)	Absent/peripheral/central, depending on the courseware and the methodology	Peripheral/central, depending on the task
The role of the technology (primarily)	CALL tutor—the program presents information, asks questions, evaluates, and gives feedback	CALL tool—the program facilitates communication or access to information

ing environment involved, but they can also be context sensitive in terms of the goals of the evaluation, the priorities invoked, and the questions that need to be asked. This brings us to reflect carefully on two points: from one direction, the value of general-purpose evaluation frameworks when the goal is to answer specific questions related to the design of the CALL materials; and from the other direction, the value of general-purpose software reviews when the particularities of the target audience are unknown, a point we discussed earlier.

The more general frameworks offer a principled approach to courseware, Web site, and task evaluation. Although somewhat involved, they provide the evaluator with (a) a procedure or process to follow and a network of elements to consider and weigh before reaching a judgment, and (b) a more powerful instrument than the early checklist approach.

The more narrowly formulated evaluation studies, responding to the diversity of CALL activity and the particular interests of designer-evaluators, are incrementally moving the CALL community toward a better understanding of the workings of programs, Web sites, and CALL tasks as learners engage with them. We have gone from looking at programs entirely out of context, merely guessing about how students might actually make use of them, to a more detailed and precise approach to evaluating what we do. This is a very positive step forward for evaluation in CALL.

Chapter 4

Computer-Mediated Communication

The greater availability of communication tools such as chat, e-mail, and conferencing programs has substantially increased the use of computer-mediated communication (CMC) in language teaching and learning. CALL practitioners have enthusiastically embraced these new technologies. As a point of departure, Herring (1996) provided a useful definition: CMC is *"communication* that takes place between human beings via the instrumentality of *computers"* (p. 1, emphasis added). This simple description can disguise the complexity of CMC. Although there are many aspects of CMC communication and face-to-face communication that are more or less shared, there are also subtle differences. Placing the computer between the communicators affects the way in which a message is composed, edited, accessed, read, and responded to. These differences are beginning to become noticeable in the fast-growing volume of research involving CMC and CALL. Any discussion of CMC needs to take into consideration the effects of the computer on the communication that occurs through it, as well as on the communication partners. It should be noted that, for the most part, text still remains dominant as a medium of CMC, but other formats are becoming more common with developments in technology.

In broad terms, CMC may be categorized as either *synchronous* or *asynchronous*. Synchronous CMC allows for an active exchange of information virtually in real time where participants can read or listen to messages and respond immediately. The downside is that all participants must be online at the same time, which can be difficult if there are differences in class times or time zones. Alternatively, in asynchronous CMC, participants can log onto the computer when it is convenient for them to do so, providing more freedom in the way that online and collaborative tasks can be implemented. Another way in which CMC may be categorized is with regard to the number of participants involved in the communication. CMC interactions do not occur only in one-to-one settings, and it is possible for many people to be involved at the same time. An example of this might be when a message may be sent from an individual to multiple recipients. This type of CMC has very different characteristics from one-to-one communications. On the one hand, it gives participants exposure to target language input from a number of different people; on the other hand, the communication no longer tends to be private, and as such participants would be expected to exercise a greater degree of caution in what they contribute to the forums. In contrast, in one-to-one CMC, it would be expected that participants would be more candid and open with their partners with less inhibition than might be seen in communication with multiple participants (see Söntgens, 1999).

Synchronous CMC includes chatting (Smith, 2003; Tudini, 2004; Xie, 2002), classroom discussions (Beauvois, 1995, 1997; Salaberry, 2000a), and MOOs[1] (Shield, 2003; Svensson, 2003). Synchronous CMC has been described as being at the "most interactive end of the CMC spectrum" (Paramskas, 1999, p. 17). The most common form of synchronous CMC is the chat room. Chat rooms can either be open or closed; that is, they can be accessed by anyone with access to the Internet, or teachers can set up private chat rooms exclusively for their own students. In their closed form, chat rooms can operate with as few as two participants in a one-to-one arrangement—there are several readily available applications that can achieve this, such as *MSN Messenger* or *Yahoo! Chat.* Although these applications can allow for more than two participants at one time, their membership is closed, and participants can enter a chat session in progress only if they are invited into it. Teachers may set up closed chat rooms for an entire class, as is often done in classroom discussions, in which all of the members of the class can contribute to the chat session. An alternative is to allow learners access to an open chat room, which anyone with access to an Internet connection can join. One of the most commonly used open chat rooms—IRC, or internet relay chat—has been the focus of several studies (e.g., Hudson & Bruckman, 2002; Xie, 2002). Another type of synchronous CMC is MOO, which is a virtual online environment in which participants interact with each other and the virtual environment.

Asynchronous CMC includes mailing lists (Hoshi, 2003), bulletin boards (Stauffer, 1994), and e-mail (Itakura, 2004). Mailing lists and bulletin boards have similarities in that participants can post messages to them, and these postings can be seen by all members. One of the biggest differences is

---

[1]There are a few definitions attached to the acronym MOO, but one of the most common ones is "Multi-user domain Object Oriented."

that, in a bulletin board system, the messages are usually accessed from and posted through a Web page according to threads, whereas in a mailing list, the messages are sent out to each member's e-mail address. Many bulletin boards may also be open to the public, whereas mailing lists are limited to the subscribers only. There are other differences between these types of communication as well, but these are dealt with later in the chapter. The most widely used asynchronous one-to-one mode of CMC is e-mail. E-mail interactions in a language-learning environment can be monolingual, in which all the interactions occur in a language chosen by the teachers, or it may be bilingual, in which both languages are used. A popular type of bilingual e-mail interaction arrangement is called "tandem e-mail," in which both participants try to maintain an even balance of both languages (see chap. 2 for a more detailed discussion).

In the description part of this chapter, we discuss recent research in the field of CMC, looking specifically at the different types from the perspective of their current applications and the focus of the research surrounding them. From there, the chapter introduces and discusses the various modes of CMC for language learning, the effects that these modes have on both the language and the participants in the communication acts, and their applicability to language learning.

DESCRIPTION

E-mail/SMS

E-mail is still probably one of the most popular forms of CMC used in language teaching, mainly because it allows students to access native speakers relatively easily, without needing special equipment or needing to match up the class times for the participants involved in the exchanges. Perhaps the most commonly cited advantages of e-mail as a learning tool are that it provides access to authentic language and serves as a means of learning more about the target culture (Gray & Stockwell, 1998; Lee, 1997). E-mail is typically used in one of two ways: monolingually, in which all of the interactions take place in one language: or bilingually, in which the interactions occur in a combination of languages. Monolingual exchanges are common in NNS–NNS interactions, or when the interactions occur between the students and/or the teacher, although there are cases when monolingual interactions occur even in NS–NNS interactions. Bilingual tandem exchanges consist almost exclusively of NS–NNS interactions.

Monolingual NS–NNS key-pal arrangements do not appear frequently in the literature, although there have been a small number of studies carried out (Itakura, 2004; Saita, Harrison, & Inman, 1998). One of the biggest criticisms targeted at this type of interaction is that, in most cases, although the tasks are demanding for the nonnative speakers, they tend to be rather tedious for the native speakers (Fischer, 1998). Of course, this problem can be greatly reduced if the nonnative speakers are paired with native speakers who are focusing on some other aspect, such as culture. Although such imbalances between roles in environments where all of the participants are nonnative speakers is not an issue, other problems can occur, such as learners being afraid of receiving imperfect models of language input from their partners, and wanting to have errors in their language output corrected. Another alternative is to have the teacher respond personally to the students' e-mails, but there are difficulties here, too. Even if teachers do not attempt to correct student output in any degree of detail, the act of simply replying to each student in the class can take up enormous amounts of time, particularly with large classes (see Aitsiselmi, 1999).

An arrangement that caters to these difficulties is a tandem e-mail exchange. As described in chapter 2, tandem learning depends on the principles of reciprocity and autonomy. Learners must try to maintain a balance between what they contribute and what they receive, and also to take responsibility for their own learning. There have been a number of studies performed in tandem learning, and teachers have reported very good results, particularly from the viewpoint of student motivation (e.g., Leahy, 2001). Woodin (1997) described the benefits of tandem learning as exposure to the target language, active learning, negotiation of meaning and information seeking, error correction, use of cultural information acquired by participants, and reuse of language offered by participants' tandem partners. The primary differences between tandem learning as opposed to monolingual e-mail exchanges are the existence of error correction and the reuse of language offered by the tandem partner. Although error correction does exist to a degree in monolingual e-mail exchanges, both partners involved in tandem-learning exchanges are often encouraged to actively correct their partners' e-mails, serving to help both sides to become more aware of their own errors.

Studies into tandem learning have occurred across a number of different languages and have investigated several different aspects of this type of CMC-based learning. Appel and Gilabert (2002), for example, investigated the effects of motivation on task performance in their study of Irish and Spanish learners. Their results suggested that tasks that were of more direct relevance to the learners were more motivating and increased both the number of messages and the number of words in the e-mail exchanges. In another study between British and German learners, Leahy (2001) posited that learners exhibited improvements in their vocabulary range and in reading all the subject specific material in the target language as a result of the tandem exchanges, and that learners also experienced an increased awareness of the cultural differences between the participants. A commonly used forum for e-mail tandem learning is the International E-mail Tandem Network (Appel & Mullen, 2002; Little & Ushioda, 1998; Woodin, 1997), which enables learners to locate partners for tandem e-mail exchanges.

In recent years, e-mail itself has taken on a new form as a result of the introduction and capabilities of mobile devices, including PDAs and, of course, the mobile phone. The features that are available on each of these devices vary greatly from country to country. In some countries, these mobile devices can operate essentially as miniature laptop computers, whereas in others they are less sophisticated (see chap. 8 for more discussion on language-learning technologies). E-mail via the mobile phone has been available for many years in Asia, particularly Japan and South Korea, whereas it is a more recent introduction to many other areas of the world. However, a similar system called SMS (short-messaging system), has been used across much of the world. SMS shares some of the functionality of e-mail except that it is typically only available to be sent and received through mobile phones, and there are often regional restrictions under which messages can only be sent within the same country or zone of the network of the mobile phone provider. Mobile e-mail or SMS obviously provides learners with a lot of freedom as to when and where they access the e-mail; however, there are disadvantages, such as the lack of a keyboard and the small size of the screen. This will, of course, affect the quantity of information that can be exchanged between the participants, both in terms of the speed at which a message can be typed and the amount of text that can be read. It is not surprising, then, that when given the choice, learners will often opt for PC-based e-mail as opposed to mobile-based e-mail, particularly in second-language learning.

There has been a wide range of tasks that have been adopted by teachers in their e-mail language-learning environments. Whereas some teachers have provided learners with topics to discuss with their partners based on perceived areas of interest and goals of the classes (e.g., Appel & Gilabert, 2002; Stockwell & Harrington, 2003), there have also been various collaborative tasks adopted, such as that reported by Nelson and Oliver (1999), who described a project in which learners collaborated to solve a murder mystery while communicating with each other via e-mail. There have also been cases of successful e-mail projects for which topics have not been set, and learners have been free to pursue whatever areas of interest they may have. Although free e-mail discussion may be possible when the learners are highly motivated, in the majority of cases the e-mail exchanges quickly lose momentum after the first few messages if concrete goals are not set for learners. Thus, although the responsibility lies with the learners to maintain the e-mail interactions, there is also responsibility on the part of the teacher to ensure that the tasks set are sufficient to hold learner interest.

Chat

Synchronous chat is characteristically very different from e-mail. One of the most obvious differences, of course, is that in chat the learners are required to be online at the same time, which may limit its usage in cross-institutional exchanges unless times have been prearranged. In addition, the learners have far less time to edit their messages than they do in e-mail. Despite these possible shortcomings, much has been written in favor of chat. Because of the real-time interaction of chat in which participants negotiate meaning by modifying the input and output and responding to feedback, chat bears certain similarities to face-to-face conversation (Lee, 2001, 2002b). Although, of course, negotiation of meaning is apparent in other forms of CMC as well, the synchronous nature of chat may make it more conducive to negotiation than are asynchronous forms of CMC. When partners are waiting for responses in real time, learners may find it easier to ask their partner for information rather than consult other resources, such as a dictionary. Researchers have found that synchronous chat contains many of the features that may be found in face-to-face communication, such as comprehension checks, clarification requests, confirmation checks, use of the L1, self-corrections, word invention, requests, use of approximation, and communication and compensatory strategies (Blake, 2000; Lee, 2001; Smith, 2003). Despite sharing many aspects of face-to-face communication, it is argued that synchronous modes of CMC afford learners some benefits over face-to-face interaction, especially that there is some indication that the text-based medium may amplify students' attention to linguistic form (Warschauer, 1997). In addition, the ability to print out the logs of learners' interactions is a useful monitoring and assessment tool for students studying at a distance, as well as providing a "snapshot" of a learner's interlanguage (Tudini, 2003).

As described earlier, chat sessions can be either open or closed. In open chat sessions, learners have access to an almost unlimited body of native speakers with whom to interact, meaning that the learners are very unlikely to be aware of whom they will be interacting with during the chat sessions. On the downside, however, learners may also be exposed to language that is improper or inappropriate without being fully aware of the connotations or nuances of this language (see Levy, 2006). Closed chat rooms most frequently consist of only the learners, or the learners and the teacher, of a single class. This arrangement allows for a lot more control over what is discussed in the sessions, but the input is limited to participants who are generally known to all.

Chat sessions, whether open or closed, obviously have fundamental differences from face-to-face communication, the most obvious being that all chat-based communication, be it text-based or voice, is non-visual. As with e-mail, this means that the lack of visual clues must in some way be compensated for by more explicit means, again either in the text somehow, or through speech. Although it is impossible to capture textually all of the facial expressions that are evident in face-to-face communication participants in chat sessions will often use what are called "emoticons" (emotional icons), which are discussed later in the chapter. Another difference between chat and face-to-face interaction is that the communication that occurs in chat is often disrupted and discontinuous, with many different topic strands and interactions being carried out simultaneously (Negretti, 1999). The burden on the learners is immediately obvious. Not only do they need to familiarize themselves with the target language, but also with the communicative differences of the chat environment. In addition, the anonymity that chat allows can have both positive and negative effects. From a positive viewpoint, the fact that the identities of the learners are hidden means that some of the shyer learners may be more willing to participate in the sessions than they might be in face-to-face situations. On the negative side, however, this anonymity often gives rise to a phenomenon known as "flaming" (discussed in more detail in the discussion later in the chapter). It is not surprising, then, that chat possesses a different social dimension from what might be found in face-to-face communication. Although there are cases in which the identity of all participants in a chat session are known, open chat often tends to be anonymous, contrasting with other forms of CMC such as e-mail, in which the identity of the message author is generally clear. As Paramskas (1999) described, this anonymous nature often results in difficulties related to tracking participants and in blurred realities. Furthermore, as Darhower (2002) found, during chat sessions some learners assume an identity quite different from their real identity, often taking on different personas or genders.

The tools used to facilitate chat are varied, and include *ChatNet IRC* (Smith, 2003); the "virtual classroom" component of *BlackBoard*, (Lee, 2004); *Open Transport Chat* (OT Chat; see, e.g., Fernandez-Garcia & Martinez-Arbelaiz, 2002); *Remote Technical Assistance* (RTA), which is a part of Windows software (Blake, 2000); *Aspects* (Salaberry, 2000a); and *WebCT* (Darhower, 2002). The choice of chat tool may depend on the language of the chat sessions, either for technical or practical reasons. Xie (2002), for example, adopted an IRC program called *mIRC* as a part of teaching his Chinese classes, because it allowed the participants to input and read Chinese characters, something that was not possible in many other IRC programs. Tudini (2003) required her students of Italian to participate in the chat site entitled "C6," to provide access to native speakers that might not be possible through other chat systems. There has also been an increasing number of studies into synchronous chat conducted in MOOs. Although, of course, there are several similarities between the chat in MOOs and through other means, there are also differences as well, as is described in the next section on MOOs.

Choice of task is also an important issue in chat, and many researchers have advocated setting topics or tasks for learners before starting the chat activities (Lee, 2002a). As with e-mail, chat sessions without focus tend to be quite brief and superficial in content, so setting the topics encourages more goal-oriented, cross-cultural, useful, and lengthy discussions (Tudini, 2003). Because participants in the chat sessions are waiting for responses from their partners in real time, learners are less likely to drop off their participation in the chat sessions as they might do in asynchronous modes of CMC, such as e-mail. However, taking the time to select the task around which the chat sessions are centered will ensure a more focused and meaningful discussion.

MOOs

MOOs are virtual environments, in which participants can meet together and interact with each other and the environment. MOOs have been described as social worlds or "non-physical spaces for language learning" (Svensson, 2003, p. 125). Not originally designed for language learning, MOOs originated as a type of online gaming system. One of the original MOOs designed specifically for language learning is *SchMOOze University*, in which learners enter a virtual university campus, moving around the campus grounds and interacting with other "students" at the university. Dedicated language-learning MOOs have also been developed for other languages, such as French (*LeMOOFrançais*), German (*Dreistadt MOO*), and Spanish (*Mundo Hispano*), to name a few.

Typically text based, MOOs require participants to read the written descriptions of their environments. As described earlier, MOOs have some features in common with chat, but there are also several aspects in which they differ. Whereas in chat rooms participants are only interacting with other participants, in MOOs participants are also able to look at, interact with, and even change their environment. Another difference is that although MOOs are a synchronous form of communication, they also have the potential to be asynchronous in that members can send MOOmail—"a type of in-MOO e-mail system" (Shield, 2003, p. 106)—and participants can also leave notes for other participants. In addition, whereas in most chat rooms any messages typed are automatically sent to all participants in the chat room, in MOOs participants have the option to communicate only with the other people within a single room, to whisper messages to a single person, or to yell messages to everyone in the MOO. Although there are extensive similarities between MOOs and chat, perhaps one way to distinguish them is to say that the MOO contains an extra dimension that is not present in chat: imagination. Like chat, MOOs provide learners with access to authentic communication and content, but in addition have the potential to trigger learners' creativity (Emde, Schneider, & Kötter, 2001) and to foster exploratory self-directed learning (Peterson, 2001). It is not surprising that MOOs have attracted the attention of language teachers. Within the "walls" of the MOO, teachers can place students in virtually solely target-language environments, making it possible not only to learn language, but also to teleport the classroom into the target culture (Donaldson & Kötter, 1999).

MOOs themselves have evolved gradually over the past few years. It is only in recent times that MOOs have been accessible from Web browsers such as *Internet Explorer* or *Netscape* rather than using applications such as *Telnet*. They have also developed in their complexity, and now have the potential to include sounds and graphics, or hyperlinks to Web pages or multimedia files (Shield, 2003). Not all MOOs are text based. One type of MOO, *Active Worlds*, is a graphically impressive virtual world, where people can freely join into one of the worlds and interact with the people and objects in that world. In its free version, the site allows for participants to chat with each other, and to navigate around a variety of worlds that exist within the so-called "universe" created by the developers and some of the users. Some worlds allow unlimited access to anyone who visits the site, whereas others require participants to be registered members to join in. In the full registered version, participants are not only able to interact with the environment, but also to build and create objects within one or more of these worlds. Although it was originally developed in English, there are other language environments available through *Active Worlds*, including Italian, French, and Russian. MOO usage depends very much on the imagination of the teacher. Svensson (2003), for example, described an innovative method of giving presentations in this environment, where learners are required to gather together in one place in the MOO to "watch" student presentations. Toyoda and Harrison (2002) portrayed a virtual version of Nagoya University, which they developed for students of Japanese.

Despite the visual attractiveness of graphically interfaced MOOs such as *Active Worlds*, there are people who argue that text-based MOOs have pedagogical advantages (e.g., Shield, 2003). Of course, learners are exposed to more language in that they must read descriptions of "locations" in the MOOs in order to understand where they are or where they want to go. In addition, learners must textually describe their emotions and actions rather than having them visually displayed, as in graphic-based MOOs.

Research into MOOs for second-language learning has mainly focused on the chat function of the MOO, and although there are differences between the chat in MOOs and other forms of chat, these differences have received very little attention. One area that is emerging as a research area in MOOs is tandem learning (Kötter, 2003; Schwienhorst, 2002). These studies have suggested that although many of the potential benefits associated with tandem learning through e-mail are also available to learners through MOO-based environments, there are also possible pitfalls. In tandem e-mail it is relatively easy to determine the degree to which learners are using one language or the other, whereas large imbalances in the language used are apparent in MOOs (Schwienhorst & Kapec, 2003). Should these imbalances become too large, the reciprocity required for tandem learning to be successful becomes strained, and it is possible that the exchanges will not continue.

MOOs allow for a combination of communication modes (see chap. 7 for more discussion) for exploratory learning and virtual immersion in a language-learning environment, and as such have potential as a tool for second-language learning. There is, however, still a stigma attached to them that places them more in the category of "game" as opposed to "learning resource" (Shield, 2003), which may discourage some teachers from exploring their applicability in the language classroom. This perception may be perpetuated by the fact that many MOOs still use the metaphors of the gaming versions, referring to administrators as "wizards" and other users as "players" (Kötter, 2003). Still, attitudes toward the medium have shown a steady change, and more and more teachers are starting to experiment with the possibilities that MOOs can afford them.

Conferencing

Although a potentially powerful learning tool, conferencing in second-language learning has featured less in the research than have the forms of CMC listed previously. Broadly speaking, conferencing can take one of two different forms: It can include synchronous chat or it can include audio and visual messages transferred by the computer. These types of conferencing possess rather different characteristics. Although, of course, they are both synchronous, one of them is text based whereas the other is speech based. Text-based conferencing fits most comfortably into the field of synchronous chat, and as such has been covered earlier in the chapter and is not be dealt with here.

Research into video- and audioconferencing to date has been predominantly in the area of distance language education. As these new tools are adopted and adapted by language teachers, there are effects on the learning environment and the interactions between the student and the tutor (Wang & Sun, 2001). The effects of the physical differences between teacher and learner can be reduced, and teachers and learners can hear and see to whom they are talking, which gives many learners the opportunity for oral practice that was not possible through normal distance courses (Hampel & Hauck, 2004).

Traditionally, audio- and videoconferencing have been costly and re-quired special equipment. However, the spread of broadband technologies and great reductions in the costs of hardware and software have made these types of conferencing more affordable for language teachers operating even on low budgets. In distance language courses, which rely heavily on au-dio and video clips, streaming media programs such as *CUSeeMe, NetMeeting, Paltalk*, and *iVisit* provide affordable two-way video on personal computers (Godwin-Jones, 2003). Even many chat programs, such as *Yahoo!* and *Windows Messenger*, have started to provide audio and video capabilities with varying degrees of sound and picture quality (see Cziko & Park, 2003, for an overview). *Skype* is a relatively new addition to the range of audio-conferencing tools, and offers very good sound quality over broadband connections.

However, there are still difficulties associated with audio- and video-conferencing for language learning. One of the major difficulties is the clar-ity of the picture and sound available to participants. Because of the bandwidth required for sending audio and video content across the Internet, picture resolution and audio compression often must be reduced to a minimum. This can make video images jerky and difficult to see, and audio messages can be unclear with breaks or lags. As a tool for language learning, this consideration is not a minor one, and participants on both sides of the interactions need to ensure that they have sufficient hardware as well as Internet capabilities to have an intelligible message. Compared with tools such as e-mail and chat, which require no more hardware than an Internet connection and the keyboard, audio- and videoconferencing need microphones and cameras, which may take some skill to set up.

In addition, as Hampel and Hauck (2004) noted, the effective integra-tion of conferencing is complex both pedagogically and technically, and failure to plan for each of these has the potential to detract from the lan-guage-learning environment. Still, the value of conferencing in language learning is indisputable, providing a means through which learners can practice oral and aural skills even when geographically separated from their communication partners.

Mailing Lists/BBS

Other forms of CMC that have featured in second-language-learning envi-ronments are mailing lists and BBSs (bulletin board systems). Both mailing lists and bulletin boards are asynchronous systems in which messages are sent to multiple recipients. The difference between them is that messages sent to mailing lists are typically sent out to members' e-mail addresses in chronological order, whereas messages sent to bulletin boards are usually threaded, depending on the topic. Bulletin board messages may or may not also be sent as an e-mail to members' e-mail addresses.

Bulletin boards and mailing lists have two primary applications: They can be used for learners to discuss issues *in* the target language (Lamy & Goodfellow, 1999b), or alternatively they can be used to discuss issues *about* the target language (Hoshi, 2003). In both cases, although there are a number of pedagogical applications, the lack of privacy of the information may have an effect on the intimacy of detail that participants wish to post.

Bulletin boards and mailing lists are often used alongside other computer-mediated tools. Vick, Crosby, and Ashworth (2000), for example, used a combination of MOOs, e-mail, mailing lists, chat, *CUSeeMe* videoconferencing, and *CoolTalk* Internet phone, allowing for both synchronous and asynchronous communications between the teachers and the students in a project in which students produced a Web magazine and planned and developed an "ideal town". Möllering (2000) described an environment in which learners of German in an online course communicated with each other using the bulletin board and internal mail of *WebCT*. Students used the e-mail function for submitting parts of assignments and to request information (predominantly in English), whereas the bulletin board was used for communicative tasks in German. Other examples are given by Strambi and Bouvet (2003), in which CD-ROM-based tasks were coupled with the bulletin board part of *WebCT* and telephone interviews, and Sotillo (2000), in which learners combined both asynchronous (a BBS) and synchronous (classroom chat) sessions with computer-based essay writing and teacher-fronted sessions. This type of learning, which involves the use of various modes of CMC, is referred to as "multimodal learning" and is discussed in more depth in chapter 7 on practice.

If we consider the different forms of CMC described here, we can see that there are very large variations in the ways in which they are used, the language that is used between the participants involved in the interactions, the nature of the interactions, and even the approach to language learning. The following section examines these variations, identifies the applicability of the different forms of CMC according to language-learning goals, and describes what CALL practitioners need to bear in mind when making choices about which is most appropriate.

DISCUSSION

Technology, Mediation, and Communication

As noted at the beginning of this chapter, any discussion of CMC needs to recognize and include the effects of the computer on the communication that occurs through it. It is incorrect to assume that the instrument that mediates communication exerts no effect, or that all forms of mediation exert the same effect. Different technologies influence communication in different ways. Stepping back for a moment, it is useful to look at communication that is facilitated not only by the computer, but also by other types of technology, such as the telephone, a microphone, a loudspeaker, faxes, letters, or virtually any type of technology that can facilitate, enhance, or alter the way in which we communicate.

Some of these technologies have become so ingrained in our daily lives that their effects are almost forgotten. Although we pick up the telephone and have a conversation with someone in much the same way as we do in face-to-face situations, there are features of the communications that still differentiate them. Hutchby (2001), for example, discussed the roles of the caller (the person who makes the phone call), the answerer (the person who picks up the telephone at the other end), and the called (the person whom the caller intended to call). There are different expectations associated with each of these three roles. The answerer will most typically answer the telephone with "Hello?"; and use of language other than this, except in the case of close friends and family who know who the caller is, may provoke uncomfortable reactions. Similarly, there is a set order that users of the telephone follow when communicating, such as greetings and confirmation of the caller and the called, as well as ways in which to hang up the telephone. It is also well worth remembering that conventions vary from culture to culture (see Liddicoat, 2000). Added to this is the fact that the participants in telephone communication are not able to see each other, and periods of silence might be met with comments such as "Hello?" or "Are you still there?" Microphones and loudspeakers also affect the type of language used. By the nature of these technologies, the communication is typically one way, and although the speaker and the listeners may be able to hear each other, the language itself is different from what it may be without such technologies.

Written forms of communication also affect the ways in which the communication is conducted. Faxes allow for messages to be instantly delivered to our intended audience anywhere in the world, and permit use of not only textual but also graphical information. E-mail remains generally text based, but various types of multimedia can be appended to messages with very little effort. These differences also have the potential to affect the way in which a message is communicated to a recipient.

Modal Considerations of CMC

These simple examples clearly illustrate some of the effects that the computer can have on the type of messages that are conveyed and the complexity of the issues involved. The choice of media used for the communication has the potential to affect the message in the amount of time it takes to be sent and received, the relationships between the participants in the communication, the types of language used in the message, the types of equipment necessary in order to conduct the communication, and even the preferences of the individuals involved. Fig. 4.1 shows the different types of CMC, and the modal considerations that must be considered in using them for language teaching.

Each of the different types of CMC differs in its modal considerations, and these have the potential to constrain or enable the communication that occurs through them, in a concept referred to as *affordances* (Gibson, 1979). As Hutchby (2001) argued, there is a "complex interplay between the normative structures of conversational interaction and the communicative affordances offered by different forms of technology" (p. 13). That is to say, in CMC the technology plays a major role not only in the choice of language used, but also in the types of messages that can be conveyed, the social relationships that can be formed, the psychological pressure that participants may feel, as well as the choice of tool in conducting the communication (see also Levy, 2006). The effects of each of these dimensions of CMC for language teaching are not insignificant, and are dealt with individually in the following sections.

The Temporal Dimension

As can be seen in the previous examples, different forms of communication will naturally have different temporal attributes. Face-to-face communication has, in essence, a zero time gap, and the participants in the communi-

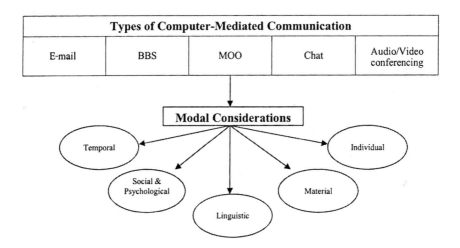

FIG. 4.1.   The modal considerations of CMC.

cation will, for the most part, expect responses to be immediate. Similar time intervals are also apparent in synchronous CMC—such as chat, MOOs, and conferencing—in which interactions also occur in real time.

As Skehan (1998) asserted, all things being equal, exerting greater time pressure on learners will mean that there is "less time for attention to form both in terms of accuracy or complexity" (p. 42). This time pressure is likely to cause more lexicalized processing, with less concern for analysis, restructuring, and accuracy. Time pressures themselves vary greatly from one form of CMC to another. Asynchronous CMC, of course, allows the learner far more time to think about a response, and provides sufficient time to consult resources such as dictionaries or grammar reference books, or even to seek assistance from other people.

Does this issue of time, then, affect the language that is used in different forms of CMC? This question can be investigated from two viewpoints—the *quantity* and the *quality* of the language produced by participants. Regarding quantity, Abrams (2003b) found there to be markedly more language produced in synchronous chat compared with asynchronous BBS, whereas Pérez (2003) discerned that although there was slightly more language produced on average in synchronous chat than through e-mail, this figure was not significant. Abrams (2003b) attributed this difference to the fact that members in the discussions were less motivated to participate due to the extended nature of the interactions, in which learners often had to wait for several days for responses. Sotillo (2000) investigated discourse functions and syntactic complexity in ESL learner output obtained via two different modes of CMC: asynchronous and synchronous discussions. Sotillo located quantitative and qualitative differences between the two kinds of discussions. Whereas students communicating synchronously seemed to focus on meaning and disregard accuracy, those communicating asynchronously had more time to plan their answers and monitor spelling and punctuation. However, malformed sentences and inaccuracies in spelling and punctuation were evident in many of the asynchronous postings. Such results need confirmation through further research. The literature on planning, for example, shows that pretask planning has the potential to significantly influence the language produced in the task that follows (see Skehan & Foster, 1997, 2001).

A term seen alongside synchronous and asynchronous is *delayed synchronous* (e.g., Hoven, 2004), which is communication that depends on the push of a key (such as the enter key) to send a message. This means that although a response may essentially be given immediately, there is still one step between composing a message and the message being forwarded to the recipient. Most chat programs would fit into this category. Chat, then, is often more than a spontaneous, unedited response as might be seen in conferencing or in face-to-face environments. As Smith and Gorsuch

(2004) explained, examination of only chat transcripts can give a distorted perspective of what happens during the message composition process in chat, because only the final product achieved directly before the enter key is pressed is available to researchers. It follows, then, that the time pressures exerted on learners will vary depending on what type of synchronous CMC is being used.

Thus, the issue of time is quite a significant one regarding CMC. Synchronous forms of CMC will generally stimulate higher language output from the learners, but place higher pressure on the learners to produce language quickly. In doing so, the focus is directed more toward achieving communicative goals than toward accuracy. Asynchronous CMC, on the other hand, allows learners time to think and to process language input while also providing time for them to edit their own language output before sending it to their partner. Although this can encourage more accurate language, the fact that learners often need to wait for responses can change the "flow" of the interactions. As Sotillo (2000) argued, both modes of CMC have different discourse features, which may be exploited for different pedagogical purposes depending on the objectives of the tasks in which learners are involved.

There are further time-related, practical matters for the language teacher to consider when using synchronous and asynchronous forms of CMC. Synchronous CMC requires that all participants be available at computer terminals at given times to enable communication to take place. When there are differences in class times, time zones, or even semester overlaps, this may not be possible. Although learners at one end may be able to work from home late at night, conversing through chat with learners in a classroom in a different country, when special equipment is needed—as is required for some forms of videoconferencing—the problem becomes difficult to overcome. In one study, participants noted that it was problematic to synchronize times that would allow them to meet their partners online (Lee, 2004). Stockwell and Levy (2001) also found that students exchanging e-mails in different countries were only able to have a 5-week overlap in which to conduct their exchanges, shortening the intended length of the exchange period. These are serious considerations when embarking on any kind of CMC project in which learners are interacting with other institutions.

The Social and Psychological Dimension

Different forms of CMC vary from one another greatly in the psychological and social impact that they have on the participants communicating through them. Doughty and Long (2003) maintained that the "choices among the numerous technological options ... need to be based, in part, on psycholinguistic considerations" (p. 50). That is to say, when making a decision about which form of CMC to apply to a given environment, it is necessary to consider the situation from the viewpoint of the learner. This is of particular importance in distance learning environments, which are predominantly asynchronous and where essentially the only contact that occurs between teachers and learners is through the computer. Indeed, a great deal of the research into distance learning focuses specifically on how the choices of CMC can ease the burden on the learners (e.g., Kötter, Shield, & Stevens, 1999; Strambi & Bouvet, 2003), and how to "compensate for the asynchronicity of communication and the lack of proximity between instructor and learners" (Doughty & Long, 2003, p. 51). This need for support for the learners in a distance setting is related to the concept of *social presence* in CMC, defined by Wood and Smith (2005) as the degree to which we perceive others in an interaction as "real" people. Presence varies greatly depending on the medium. Of all of the types of CMC described in this chapter, videoconferencing would provide the greatest sense of presence, because the participants in the interaction are able to see and hear each other. Synchronous forms of CMC will obviously offer a greater presence than asynchronous ones, in that responses are received in real time. This presence decreases greatly in e-mail, for which there may be a significant time lag between the sender sending the message and the receiver receiving it. Mailing lists and bulletin boards provide an even more reduced sense of presence, in that the messages are usually intended for a multiple audience, and the personal nature of the communication becomes somewhat watered down.

It is clear that CMC is socially very different from face-to-face communication. Another obvious example of this is turn taking. When a person begins to speak in a face-to-face environment, usually the other participants will stop and give the speaker a chance to say what he or she would like to say. In the majority of chat and other synchronous environments, however, other participants are not aware when one person starts to type a message and may continue with a topic, or else may change the direction of the discussion while a potential contributor to the discussion types his or her message. This often leads to a situation in which one person's contribution to a discussion is often out of step with what has gone before it, even though when that person started to write the message it was still in sync. Thus, when examining synchronous CMC exchanges, one will often see topics moving in a rather cyclical manner as opposed to completely linearly—the topic moves in one general direction but often backtracks as out-of-step contributions appear. As Negretti (1999) showed, the responses to questions are often quite delayed, and topic strands are frequently intermingled. The characteristics of asynchronous CMC are different altogether again. As described by Stockwell (2003a), messages will often contain multiple topics, meaning that there is a need for participants to flag which topics they are responding to when replying to messages. There are other differnces as well. In synchronous forms of communication, responses are immediate, and it is easy to maintain a train of thought through a series of messages. In asynchronous CMC, however, participants in the communication often need to wait several days for a response, which can "interrupt the discursive momentum" of the interactions (Abrams, 2003a).

Communication breakdown is another social difficulty associated with CMC. Whereas in face-to-face environments any breakdowns in communication can be immediately obvious due to nonlinguistic cues, they may not be as immediately obvious in computer-mediated environments. In face-to-face interactions, breakdowns in communication are generally dealt with quickly in order for the interactions to continue. In contrast, examination of computer-mediated interactions suggests that learners will often fail to deal with difficulties in communication (Lamy & Goodfellow, 1999a; Stockwell, 2003a). In fact, Stockwell (2004) showed that, when confronted with communication breakdown in e-mail interactions with native speakers, 64% of learners did not deal with the breakdown, choosing instead to ignore it. Thus, although CMC has been argued to reduce the affective filter, it should be noted that it does not abolish it completely. Evidence for this is also seen in chat. Tudini (2003) pointed out that learners of Italian involved in chat sessions with native speakers were often likely to reveal their learner status within the first 10 turns of the chat conversation. This behavior was no doubt linked to the lack of confidence on the part of the learners, and was used in an attempt to seek understanding from the native speakers for any mishaps in communication that may have occurred. In response to this, Tudini (2004) noted that the Italian native speakers in the chat sessions were likely to adapt their language to the learners' levels, including avoiding regional variations of Italian, colloquialisms, and chat jargon. Similar observations have been identified in MOOs (Donaldson & Kötter, 1999).

As described in the description section earlier, communication tools such as chat can provide learners with anonymity that enables them to communicate with less inhibition. Unfortunately, this also brings with it potential pitfalls. Hawisher and Selfe (1998), for example, warned of the possibility of *flaming* in CMC, which is characterized as bold, offensive, emotionally laden comments that are not appropriate for classroom settings. When learners are in an environment where there is some degree of anonymity, they become less accountable for their actions, possibly resulting in such bad behavior (Paramskas, 1999). Related to this is the issue of online personalities. Participants will often adopt personas in online situations that differ from reality in personality or in gender, or even will impersonate others (Amichai-Hamburger, 2005). A common reason cited for this is that women do not like to be approached by men when entering chat rooms, and hence will opt for a male or neutral persona to avoid attracting attention (Hall, 1996). Many participants will also choose not to participate in discussions, preferring to remain silent, and reading online discussions without contributing to them in a phenomenon known as *lurking* (Kollock & Smith, 1996). The reasons for lurking are varied. In discussion groups, some members may be interested enough in the content to follow the lines of discussion, but not feel sufficiently strong about the content to make a contribution. In a second-language-learning context, a more common reason is that learners lack the linguistic skills or the confidence to join into a discussion, or are unable to find the appropriate moment to participate. All of these issues are of importance to educators who wish to include their students in CMC forums, whether open or closed.

As described earlier, the expression of emotions in CMC is also very different from face-to-face situations, which often occurs through the use of emoticons. Marvin (1995) provided an excellent description:

> Many smileys, and the spelled out gestures of "smile" or "grin" (emotes) are appended to statements which are not ironic or ambiguous. They are friendly gestures, indications of approval or appreciation, much as smiles are often intended in face-to-face interaction. However, smiles in face-to-face contexts can be strategic or spontaneous and unintentional. In the context of the MOO, whether expressed with the iconic :-) or the symbolic "smile", every smile must be consciously indicated. In private something flowing across the computer screen might cause a participant to spontaneously smile, but a conscious choice must be made to type it out; a participant might frown at the keyboard ... [but] decide to type a strategic smile.

Thus, even something as spontaneous as a smile or a frown in face-to-face contexts becomes conceptually different in CMC. Whereas a person may consciously hide his or her emotions in face-to-face communication, in CMC there is an added dimension in that the decision to actually express emotion is a conscious one.

The Linguistic Dimension

Within the different forms of written CMC (audio- and videoconferencing bear close resemblance to telephone conversations and are not included in the discussion here), there appears a variety of language that, although bearing some resemblance to spoken or written language, also consists of features that are very much characteristic of the computer-mediated medium itself (see Collot & Belmore, 1996, for a discussion). As described earlier, there are differences in the quantity and quality of language that is produced in both synchronous and asynchronous forms of CMC, with synchronous CMC sharing many of the features of oral discourse, and asynchronous CMC approaching written forms of language (González-Bueno & Pérez, 2000; Sotillo, 2000).

Several researchers have made attempts to compare CMC with face-to-face interactions. Warschauer (1997), for example, found that students used language that was lexically and syntactically more complex in the electronic discussions compared with face-to-face environments. Other researchers have investigated the interactional features and the modification devices used by learners in online synchronous chat, and noted that many features of the interactions resembled face-to-face interactions such as the existence of requests, clarification checks and self-correction, intersubjectivity, off-task discussion, greetings, and leave takings (Darhower, 2002; Lee, 2002b). There has also been evidence of many features in CMC that are not present in face-to-face interactions or in other written media. Emoticons are of course one obvious example, but forms of CMC such as chat also use a range of abbreviations for commonly used expressions, such as "btw" (by the way) and "lol" (laugh out loud).

There are other linguistic considerations that must be kept in mind. As Lamy and Goodfellow (1999a) argued, although synchronous CMC allows for language production that resembles speech in many aspects, it is questionable whether it is appropriate for language learning. An examination of the majority of language produced in chat sessions, particularly those that involve NS–NS interactions, shows that the language produced is highly fragmented, abbreviated, and frequently contains spelling errors. Observations of the language produced by language learners who have participated in open chat sessions reveal that similar examples of this type of language are evident in their postings, too. Chat sessions usually contain quick and short interactions, and topics may be lost in the jumble of interactions. Problems are also apparent in asynchronous CMC. In e-mail, for example, messages often contain multiple topics, sometimes making it difficult for learners to deal with the large amount of information (Stockwell, 2003a).

Despite these shortcomings, researchers argue that CMC does have a place in the second-language classroom. An area that has attracted considerable attention from researchers into CMC is its ability to provide learners with a forum through which to negotiate meaning. Negotiation of meaning is cited widely as contributing significantly to the second-language-acquisition process, and as such is given as one of the grounds for using CMC in the second-language classroom. As Skehan (1998) asserted, the negotiation of meaning that occurs through interactions is seen as an ideal mechanism through which learners can identify where their interlanguage is limited and needs to be further extended; as well, they can receive feedback at precisely the point "when it will be most useful since the learner will be particularly sensitive to the cues provided to enable new meanings to be encoded" (p. 20). Studies of CMC have shown that negotiation of meaning and collaboration are common in the interactions. Students demonstrated tendencies to correct their own errors when encountering the correct form from their teacher during in-class interactions (Beauvois, 1998). Similarly, Stockwell (2003b) described cases of clarification requests on the part of both parties involved in NS–NNS e-mail interactions, as well as correction of lexical and pragmatic error in response to both direct and indirect feedback. There were, however, large differences among the learners in this regard, and it was evident that some learners were far better than others at noticing discrepancies between their own output and the input provided by the native speaker models. Similar examples of negotiation of meaning in chat sessions have been cited by Kitade (2000), Toyoda and Harrison (2002), and Tudini (2003), although it is interesting to note that the majority of negotiation of meaning in chat sessions appears to be of a lexical nature rather than to support the acquisition of grammar, which is where it has received the most attention in SLA research (see chap. 5 for further discussion of negotiation of meaning).

This leads to the question of whether or not CMC actually can contribute to second-language acquisition. Johnson (2002) argued that much of the justification for introducing CMC into language courses is based on the *potential* of the technology to contribute to learning, rather than on empirical evidence. Preliminary evidence suggests that it can help, and there is a growing body of research indicating that a relationship between CMC and SLA may exist. Stockwell and Harrington (2003), for example, posited that learners involved in e-mail interactions demonstrated increases in both the accuracy and complexity of the language produced, according to T-unit (minimal terminal unit) measures. Similarly, Salaberry (2000a) provided evidence that Spanish learners could develop morphosyntactic skills through synchronous online conferencing, and Payne and Whitney (2002) found that learners showed significant increases in oral proficiency when involved in online chat sessions. Other studies have been less conclusive. Gonzáles-Bueno and Pérez (2000) and Abrams (2003b) argued that although learners increased the amount of language produced and improved their attitudes toward practicing the target language, there were not significant increases lexically or syntactically in synchronous or asynchronous CMC discussions when compared with face-to-face environments.

One of the major differences between the studies that demonstrated increases in second-language proficiency and those that did not was the degree to which the learners participated in the CMC sessions. In the study by Gonzáles-Bueno and Pérez (2000), students contributed only nine messages each, and in the study by Abrams (2003b), the students participated in only three sessions throughout the semester. On the other hand, Payne and Whitney's (2002) students participated in 21 sessions over a 15-week period, and in the example cited by Stockwell and Harrington (2003), students only started to show improvements in their language output after a minimum of 10 messages. Sufficient participation by the students involved in CMC, then, appears to be an important condition for development of second-language proficiency.

The Material Dimension

The material dimension concerns the actual physical properties of the technology, such as its size, weight, and portability. It also covers the ways the technology provides for human–computer interactivity, such as the size of the screen and the keyboard, the number and size of keys, provision of a mouse or pad, the range of interactive modes on offer (text, image, video, voice), and so forth. These material dimensions (or affordances) constrain, enable, and shape the communication that occurs, although these effects are subtle and difficult to pin down with precision. One good example is the mobile phone whose physical properties (i.e., the small screen and keypad) contribute to the ways in which users communicate. The portability and connectivity of the mobile phone are defining attributes that enable users to communicate more frequently and at times and places that are optimally convenient; users do not have to seek a public phone or wait until they return home, as they formerly had to do. The smallness of the screen and keypad also lead users to a new, abbreviated variety of language (see Crystal, 2001). These technological aspects are, of course, not the only ones to influence communication. In parallel, new social and cultural behaviors and communities grow up around each new technology, such as the mobile phone, and these aspects combine with the material properties of the technology to shape the ways in which it is used.

Communication technologies used in the wider world are not necessarily appropriate or relevant for language learning. There is no reason why the choices that users make in the wider world should dictate or prescribe the technologies used within the educational institution. Still, clearly there are advantages if there is a reasonable degree of synchrony between the technologies the students use inside and outside the institution (horizontal integration). Having a degree of correspondence, when the situation allows it, offers the distinct advantage that less time and attention will need to be allocated to learner training in the institution. Students will already be familiar with the technologies involved and will bring that expertise to the classroom. All the same, in spite of the potential advantages of this connection, widespread acceptance and use of new communication technologies in the wider world does not necessarily point to effectiveness or value in the educational context. Effective transfer depends to a large degree on the nature of the particular technology in question and its strengths and limitations, both as a technology and as a pedagogical tool. Therefore, careful evaluation needs to occur before new technologies that have been accepted in the wider world can be welcomed into formal, educational contexts. Teachers have a right to be skeptical until a convincing case has been made regarding the value of the new technology for educational purposes.

One study that moved in the direction of mobile technologies for language learning was described by Levy and Kennedy (2005). This work built on the initiatives of Houser, Thornton, Yokoi, and Yasuda (2001) and Dias (2002a, 2002b) and the research of Nation (2001) on vocabulary learning. It applied these ideas to the learning of Italian within an Australian university context, specifically a course on Italian literature and society. A particular focus in this study concerned the timing and the number of repeated messages, and the nature of the recall prompts to enable effective vocabulary learning. Other functions were tested too, especially those that were consistent with the informal, social, and entertainment functions of mobile phones in the world outside the educational institution. Ideas that were explored alongside the main theme of vocabulary learning involved providing course reminders, related Internet sites, quizzes, information messages, proverbs, translations (e.g., idioms), and questions to do with song titles, first lines, and authorship. Eighteen students took part in the study, and participants were surveyed through a questionnaire and focus-group interview to gauge attitudes and expectations toward mobile phone technologies in an educational setting. Responses were very positive, although any conclusions need to be tentative because only a small group of motivated students were involved.

The Individual Dimension

Individuals have personal preferences with regard to the media they choose. This choice may depend on who the recipient of a message is, or on the type of communication employed. For example, a handwritten letter may be thought to be more personal than a typed one, and the telephone allows us to hear the voice of the person with whom we are communicating. As individuals, we have our own preferred means of communication for certain types of messages to certain recipients, and we may choose to write or type a letter, or to use the telephone to convey our message. The same is also applicable in language-learning contexts. The classes that we teach are made up of students with different background knowledge, different skills, and different likes and dislikes, and research into these individual differences has attracted an increasing amount of attention in recent years (e.g., see Ehrman, Leaver, & Oxford, 2003; Robinson, 2002).

It is not surprising, then, that when involved in communication in the target language, learners will prefer to use one type of technology above another, depending on the language-learning situation. Evidence of this may be seen in a study by Pérez (2003), who surveyed 24 learners of Spanish involved in interactions with the teacher through both e-mail journals and synchronous chat regarding their preference between the two means of communication. Pérez found that 50% of the learners preferred e-mail and 50% preferred chat. This result is very informative regarding the variation that may be seen in a group of learners, because it showed that what one learner prefers to use may be very different from the preference of another. In addition, Johanyak (1997) demonstrated that even the language that is used within a particular form of CMC will vary as a result of the background and preferences of the learner. As described earlier, although there is a tendency for different forms of CMC to resemble written or spoken language, the choice of genre used by learners will depend on their social, cognitive, and contextual experiences, in many cases irrespective of which form of CMC the learners are using.

CONCLUSION

As we have described in this chapter, there are several forms of CMC, each exhibiting different features that affect their use in language learning. For example, there are distinct differences in the time for processing language input with synchronous and asynchronous forms of CMC, and each has different social, psychological, and material considerations. To this add the variations of language used in different forms of CMC as well as the individual preferences of learners, and it is clear that choosing the correct form of CMC for a language-learning context is not straightforward. Decisions need to be founded on knowledge of the constraints and characteristics of the CMC tool itself, together with a pedagogical understanding of how the communication technology might be most effectively used for language learning. Successful implementation of CMC modes in CALL depends on having clear pedagogical objectives in mind, knowledge of the technological options and an awareness of the needs, goals, and skills of the learners (see chap. 7). Selecting the right form of CMC must build on these principles, in addition to knowing the modal considerations of each form of CMC.

Researchers have argued that synchronous and asynchronous CMC bring with them certain characteristics that make them more suitable for certain situations (e.g., Sotillo, 2000). As Skehan (1998) described, synchronous CMC places a higher cognitive load on the learner, and as such is better suited to higher-proficiency learners. Asynchronous CMC gives learners more time to process and produce input, and may be thought to be suitable for lower-proficiency audiences as well as higher. Synchronous CMC normally lends itself to greater output on the part of the learners, but this is often at the expense of accuracy. Asynchronous CMC provides learners with an opportunity to produce planned, edited language output and receive substantial generally well-formed input, but the fact that there are often time lags between sending and receiving messages may reduce learner motivation. The social and psychological aspects of communicating through the computer are also very important. The conventions of turn taking vary greatly depending on the mode of communication, with multiple topics being addressed simultaneously in e-mail interactions (e.g., Stockwell, 2003a), or intermingled topic strands between several participants, as is often the case in chat (Negretti, 1999). This may place a heavier cognitive load on learners who are not familiar with the format of these types of interactions, and thus students would need support in learning how to deal with new and different formats. However, the number of students who come into the language classroom with knowledge of communication tools such as chat is increasing, and as such this is becoming less of a problem. It is here that knowledge of the learners experience is essential in determining how the students will be able to handle the medium in their language learning.

A key consideration when using CMC for language learning concerns the language students use. The language of audio- and videoconferencing is very close to what learners may encounter in face-to-face interactions or through the telephone, so there is less concern regarding exposure to language that is ill-formed or broken. In contrast, however, language produced in other communication modes such as chat and MOOs is, in many cases, very unlikely to be considered as good models of target language input. Does this mean, then, that chat and MOOs—and to a lesser degree e-mail—are not appropriate for second-language learning? Evidence from the literature suggests that this is not the case. First, apart from the substantial motivational benefits (e.g., Coniam & Wong, 2004), although the type of language input may not necessarily be well formed, provided that learners are made aware of the fact that the language of chat is applicable only in chat environments and not in other forms of written or spoken interactions, learners still have the opportunity for development in the target language socially, lexically, and syntactically (Toyoda & Harrison, 2002; Tudini, 2003; Yuan, 2003).

Although many types of CMC commonly used in classroom environments, such as e-mail or chat, do not require any specific equipment or software that cannot be obtained either for free or at relatively low cost, more sophisticated systems, including audio- and videoconferencing, do need to have certain hardware and sufficient bandwidth to allow interactions to take place. This also means that adequate support for this hardware must also be available to deal with any problems that may arise. If learners are expected to take part in CMC-based tasks outside of the classroom, then it is important to confirm that they have access to the necessary hardware and software, as well as enough technical ability to set up and operate the tools they are required to use.

Finally, it is important to realize that learners are individuals with their own views about different types of communication media. For this reason, providing learners with choices that will allow them to follow their own individual preferences is advantageous when it can be arranged. This may contribute to a higher degree of participation and sustainability in an activity, which in turn provides greater opportunity for target-language development (Stockwell & Levy, 2001). Further discussion regarding the use of CMC and other technologies for teaching specific language skills and areas is included in chapter 7.

Chapter 5

Theory

Those working with technology in language learning—be they designers, language teachers, or researchers—want their approach to their work to be principled; that is, they seek a sound basis or foundation for the decisions they make and the directions they follow.[1] It is with this goal in mind that researchers turn to theory: They wish to build on previous knowledge by rigorously testing current theories with a view to refining and improving them (Jordan, 2004). It can also be the reason that designers and language teachers turn to theory, although their goal is a rather different one. In a nutshell, what theory provides is a position from which to view a problem. Through this positioning, theory limits our field of view, highlights what we focus on, and—most important—what we do not. It also provides an inferential structure for analysis and interpretation. CALL practitioners turn to theory, or a theoretical framework or model, to provide them with a point of departure or a sound basis or foundation. For example, we saw earlier in chapter 2 how Strambi and Bouvet (2003) used theory to inform their design. For these *designers*, a social-constructivist approach was the point of departure, one that allowed for the coherent integration of affective, cognitive, and interactionist perspectives. The theory helped them develop a view of learning and of learners that, in turn, shaped the design and development of two distance language courses for beginners in French and Italian. For *language teachers*, Egbert and Hanson-Smith (1999) also described a model for CALL based on a theoretical framework that "a teacher can use to guide [his or her] deployment of technology in the language classroom" (p.

---

[1]CALL designers, language teachers, and researchers are considered separately in this chapter, even though clearly the roles may overlap. The separation is helpful with regard to theory because of the different ways each group tends to view theory and make use of it.

3). It is worth noting that it is as a "guide" that theory is used for language teaching, and not as a prescription. Then, as *researchers*, Fernández-Garcia and Martínez-Arbelaiz (2002) appealed to a theoretical model to accomplish two goals: first, to describe the negotiation routines in an online chat environment; and second, to guide the data collection such that it may be used to provide grounds for determining whether the online chat interaction has led to language acquisition. Thus, the theory helps the researcher to decide not only what data to focus on and collect, but also how to interpret this data once it has been collected.

In general, then, theory assists CALL practitioners in a variety of ways, be they designers, language teachers, or researchers. Neuman (2003) provided a neat summary of the value of theory in a little more detail:

> Theory frames how we look at and think about a topic. It gives us concepts, provides basic assumptions, directs us to the important questions, and suggests ways for us to make sense of data. Theory enables us to connect a single study to the immense base of knowledge to which other researchers contribute. To use an analogy, theory helps a researcher see the forest instead of just a single tree. Theory increases a researcher's awareness of interconnections and of the broader significance of data. (p. 65)

The first sentence in this extract deserves careful attention. By framing what we see, theory inevitably includes and excludes, promotes and demotes. Factors that some theories consider vitally important may be ignored or consigned to a background role by other theories. In other words, every theory brings certain ideas, issues, or constructs into the foreground and consequently, by design or by default, pushes others into the background. For example, among learning theories, cognitive theory places a very strong focus on the learning processes of the individual, whereas sociocultural theory brings the social aspects of language learning much more into the foreground. It is not that cognitive theory specifically denies the role of the social, or that sociocultural theory denies the role of the individual in learning; rather, it is more a matter of where the priorities are placed, and the territory over which the theory may be effectively applied.

Also note that Neuman (2003) addressed theory in relation to the researcher. Theory is often discussed in this way, as if it is used only by the researcher. However, in CALL, as in many other practice-oriented areas of education, this is not the case. Theory is also used by designers and language teachers, a point that tends to be overlooked. The designer and teacher perspectives on theory are most important, and the nature of these perspectives, and the way they differ from the researchers' understanding and use of theory, are a key point of focus through this chapter.

Another important point that must be raised at this early stage of the chapter concerns the number of theories with possible application in

CALL. There are many theories potentially available for practitioners to draw on in their work. The question of whether this theoretical "pluralism" is good or bad has been the subject of much debate in the literature (see Beretta, 1991; Block, 1996; Gregg, 1993, 2000; Jordan, 2004; Lantolf, 1996). For our purposes here, suffice it to say we concur with Mitchell and Myles (2004) when they concluded:

> On the whole, though we accept fully the arguments for the need for cumulative programmes of research within the framework of a particular theory, we incline towards a pluralist view of second language learning theorising. In any case, it is obvious that students entering the field today need a broad introduction to a range of theoretical positions .... (p. 2)

Hence, this chapter endeavors to describe a number of theories that have been used in CALL. These theories are drawn in the main from second language acquisition (e.g., the interaction account) and education more generally (sociocultural theory, activity theory, constructivism). These particular theories have been chosen because of their frequency of use and application in CALL. In each case, a brief description of the main attributes of the theory are provided, followed by examples of its application in CALL. There is obviously insufficient space here to do justice to each theory and to provide a comprehensive account; however, detailed references are given for those who want to follow up and read more about the particular theoretical position being highlighted.

DESCRIPTION

The Interaction Account of SLA[2]

The origin of the interaction account (IA) of SLA lies in the work of Krashen who, in the late 1970s, proposed a theoretical model of second-language learning (Krashen, 1977; see also Krashen, 1985). The model—the well-known monitor model—consisted of a set of five hypotheses: the acquisition-learning hypothesis, the monitor hypothesis, the natural order hypothesis, the input hypothesis, and the affective filter hypothesis. It was the fourth of these hypotheses, the input hypothesis, that prompted the further theoretical elaboration and empirical research that led to the interaction account of SLA. It was recognized that input alone was not sufficient for lan-

---

[2]Long (1996) explicitly rejected the use of the term *theory* in relation to the interactionist position on SLA because he maintained that the research findings that lie behind it are not sufficiently robust. He said of this perspective that it involves "a mix of well and less established L1 and L2 acquisition research findings, some rather high inference interpretation, and some speculation" (1996, p. 453). In line with this view, we refer in this chapter to the *interaction account* rather than *interaction* or *interactionist theory*.

guage acquisition (contrary to the assertions of the input hypothesis), and that interaction and learner output were necessary as well (Long, 1996). Long advanced the argument with his presentation of the interaction hypothesis (later reformulated in 1996), and Swain developed it further in her comprehensible output hypothesis (e.g., Long, 1983, 1996; Swain, 1985; Swain & Lapkin, 1995). More recent work on the role of attention and noticing in learning, interactional modifications and their role in acquisition, and focus on form has led to further refinements that have helped to motivate and sustain the interactionist tradition (Doughty & Williams, 1998; Gass & Varonis, 1994; Schmidt, 1990, 1994; see also Gass, 2003; Mitchell & Myles, 2004).

The interaction account emphasizes the role of face-to-face interaction in second-language development (Long, 1996). Central to this process is the *negotiation of meaning*, in which the learner and interlocutor(s) engage in an ongoing process of interactional adjustments (Pica, 1991). These adjustments serve to highlight particular linguistic and nonlinguistic features in the discourse that render the input comprehensible. This input can then be converted by the learner to intake, which is the basis for the development of proficiency in the second language. The ongoing negotiation of meaning additionally requires productive output on the part of the learner, which also contributes to development (Swain & Lapkin, 1995).

The IA and its pedagogical manifestation, Instructed SLA, have had a major influence on SLA theory development and on the direction of the research agenda of the field (Ellis, 1994; Sharwood-Smith, 1993; Spada, 1997). A number of CALL researchers have argued for IA as an appropriate foundation for CALL research too. These include Doughty (1991), Liou (1994), and Chapelle (1997, 1998a, 1999). The IA account, it is argued, provides the CALL researcher with important questions, research methods, and an explanatory framework for studying second language learning. IA/Instructed SLA identifies the "conditions under which ideal input and interactions take place" (Chapelle, 1999, p. 5), and these conditions in turn provide the field with a much-needed framework for theory and method.

The interaction account has been particularly well used as a theoretical base in CMC-based CALL (De la Fuente, 2003; González-Lloret, 2003; Hampel, 2003; Stockwell & Harrington, 2003). E-mail and chat are especially common CALL applications where the IA, or aspects of it, is regarded as having high relevance and value (see Aitsiselmi, 1999; Fernández-Garcia & Martínez-Arbelaiz, 2002; Lee, 2001). For example, with e-mail, Aitsiselmi (1999) looked at the IA in the analysis of a communicative e-mail exchange between the teacher and English learners of French. She described Krashen's monitor model as her theoretical basis, and discussed each of the five main hypotheses. She addressed some of the controversial aspects in the model, but continued to assert the value of aspects of the model for CALL, as others have done in language teaching and learning more generally. She argued that e-mail exchanges meet the requirements of the model by:

- Encouraging participation and helping students overcome shyness with the teacher in class.
- Offering genuine interpersonal communication.
- Providing access to authentic language input.
- Providing comprehensible input containing structures a little beyond the students' current level of competence.
- Occasionally allowing the teacher to draw students' attention to recurrent grammatical errors.
- Ensuring the major focus remains upon meaning rather than form.

Although e-mail activity was outside the specific parameters of the model, Aitsiselmi also emphasized its potential for enhancing the students' self-confidence in the foreign language and for learner autonomy. In all, Aitsiselmi (1999) drew on the theory for two main reasons: to justify the e-mail exchange, and to help identify the advantages and drawbacks of the activity (i.e., for evaluation purposes). These two reasons for the use of theory in relation to design, research, and practice are perhaps the most common reasons for appealing to theory in CALL.

Another example, this time using online chat, was described by Fernández-Garcia and Martínez-Arbelaiz (2002). They discussed negotiation of meaning between nonnative speakers in synchronous discussions among learners of Spanish. In discussing the theoretical framework, they focused particularly on the need for discourse moves that allow interlocutors to negotiate meaning to ensure that the message is understood. The interactional modifications that result are held to make the input more comprehensible and thereby to facilitate language learning (Gass & Varonis, 1994; Varonis & Gass, 1985). The theoretical model presented by Varonis and Gass provides a tool to locate and analyze such discourse structures.

Fernández-Garcia and Martínez-Arbelaiz (2002) found that they were able to identify negotiation routines of the kind required by the model, although the routines discovered in online chat display important differences, as well as similarities, compared with face-to-face, offline negotiation routines. For example, and not surprisingly, online options for expressing nonunderstanding tend to be more explicit, and are expressed through a written token of some kind, where a raised eyebrow or silence can convey the same message in face-to-face settings. Furthermore, the authors discovered that chat does not afford as many opportunities as expected for negotiation of meaning. Where negotiation of meaning does occur, it is chiefly in relation to the meaning of the lexical items. Hence, Fernández-Garcia and Martínez-Arbelaiz (2002) used the theory to strengthen the rationale and justification for their study. The model by Varonis and Gass (1985) offers a descriptive tool and, thereby, a means of describing the negotiation routines (the discourse) that occur in a chat environment. Also, the model provides a way of establishing whether learners have successfully acquired language or not; again, the theory plays a role in evaluation. Here, the authors were looking particularly for evidence of modified output, as this was hypothesized to be a condition for successful language acquisition.

One of the more recent developments in the interactionist tradition is the current interest in focus on form. According to Doughty and Williams (1998), this interest has come about because meaningful input and opportunities for interaction do not on their own appear to be sufficient for the ultimate development of targetlike language-proficiency levels. These authors argued that occasional pedagogical interventions that focus on the forms of the language are needed as well, although still within a framework of communicative, meaning-focused activities.

Levy and Kennedy (2004) pursued this idea in the way they structured CALL tasks for learners of Italian. Here, the theory was used to inform the precise structuring of the task, which was designed to facilitate balanced attention to form and meaning. The forms of the language were only considered within the context of the learners performing a communicative task. In other words, the tasks were still primarily communicative, meaning-focused activities, but with a provision for an occasional or incidental focus on linguistic form, motivated either by the teacher or the student. In this project the meaning-focused activities were conducted between students using audioconferencing, whereas the form-focused activities used recordings of the audioconference as a stimulus for reflection, in a technique called *stimulated reflection*.

The interaction account focuses on learning interactions that, by necessity, involve two or more people, or a person and the computer. As Mitchell and Myles (1998) observed, this tradition still essentially regards learning as a quality or accomplishment of the individual, one "who uses relatively autonomous internal mechanisms of some kind in order to exploit the varying spectrum of input data on offer in the interactive environment" (p. 122). In the next two sections we move toward theoretical perspectives that view language learning in a different way, primarily in social rather than individual terms.

Sociocultural Theory

Vygotsky (1896–1934) believed that learning resulted from social interaction rather than through isolated individual effort, and that engagement with others was a critical factor in the process (Vygotsky, 1978). In his view, learning was at first social (intermental), and only later individual (intramental). According to Lantolf (1994), Vygotsky's fundamental theoretical insight was that "higher forms of human mental activity are always and everywhere mediated by symbolic means" (p. 418). And the pre-eminent tool for mediation is language. Hence, Vygotsky paid special attention to the role that language plays in cognitive development and in mediating the learning process. Acquiring a language enables the learner to think in new ways by providing a cognitive tool for making sense of the world.

The notion of language as a cognitive tool for mediation is one of the most profound insights of Vygotsky, an idea derived from the work of Engels (Haas, 1996). Engels posited that humans interact with the environment using material tools that mediate the interactions that occur. Through the interactions, both the environment and humans are transformed. Vygotsky extended this notion to include language—especially speech, but also writing and other sign systems—as a psychological tool that provides the "mediational means by which higher psychological functions develop" (Haas, 1996, p. 14). Although Vygotsky's understanding of cognitive tools for mediation is intended to be purely metaphorical, Haas extended this idea to include technological tools as well. She argued that "Vygotsky's theory of mediation helps us see tools, signs, and technologies as ... systems that function to augment human psychological processing" (Haas, 1996, p. 17).

Vygotsky's work took a considerable time to reach the West: It was only in 1962 that the seminal work, *Thought and Language*, was translated. Since that time, his followers have developed and extended his work under the headings of *neo-Vygotskian theory, cultural psychology, communicative learning theory, and sociocultural theory*; the latter term is now perhaps the most common and is used here (see Jones & Mercer, 1993).[3]

The Role of the Teacher.   Within the Vygotskian view, the teacher is critical. The teacher's role is very much brought into the foreground in sociocultural theory—hence its high visibility in education. The teacher is regarded as an active, communicative participant in the learning process. The teacher acts as a support to help the student until the time comes when he or she is able to operate independently. As Bruner (1985) put it, the tutor functions as "a vicarious form of consciousness." In this regard, Vygotsky introduced the well-known theory of the Zone of Proximal Development

---

[3]It is worth noting that Vygotsky was concerned primarily with how children learned their first language. On the whole, he and his followers did not make claims about how adults might learn a second language. In this regard, Laurillard and Marullo (1993) provided a detailed critique of the extent to which a Vygotskian perspective can be sustained for students learning a second rather than a first language.

(ZPD), which posited that learners benefit most from tasks that are just beyond their individual capabilities. Learners are not able to complete such tasks on their own, but with the help of a more knowledgeable and experienced individual they are able to accomplish them. Thus, a role for the teacher is warranted, in helping learners overcome the gap between what they can do alone and what they can manage with the help of others. Also, although not in the original conception, the metaphor of *scaffolding* has been developed in neo-Vygotskian discussions to capture the ways in which the teacher works with the learner within the ZPD, and which is held to facilitate learning. Finally, another key term in the Vygotskian framework is the concept of *microgenesis*, which refers to the local, contextualized learning process that begins with children and continues with adults as new concepts continue to be acquired through social/interactional means.

As may be observed in the previous paragraph, sociocultural theory is responsible for the introduction of a considerable number of new concepts. In applying the theory in CALL, the literature shows that different authors tend to be drawn to specific concepts. For example, Gutiérrez (2003), in a study on collaborative CALL, set about identifying instances of microgenesis, as well as locating and examining collaborative episodes and the specificity of the computer as a mediational tool. The study is worth looking at in a little more detail, as an example of an application of Vygotskian theory in CALL.

Gutiérrez's (2003) research questions were threefold. The first question concerned the degree to which different tasks support collaborative work in the classroom; to answer this question, Gutiérrez analyzed the transcribed data by identifying and counting the number of collaborative episodes (CEs) produced by pairs of students in each of the different tasks. The second research question related to microgenesis and identifying instances of microgenetic activity during task completion; to answer this question, she examined instances of language-related episodes (LREs; occasions on which students talk about the language they are producing), and microgenetic episodes (MGEs; episodes in which the learning process can be observed). Finally, the last question looked at the relationship between microgenesis and the specificity of computer-based tasks; to answer this third question, Gutiérrez compared computer-based and paper-based tasks to reexamine study patterns and issues that provide specific information on the role of the computer (e.g., higher indexes of LREs, CEs, and MGEs).

Of particular interest here is the way in which the sociocultural theoretical framework and the various strands and key concepts within it infiltrate, shape, and frame the thinking of the researcher. The theory and the key concepts determine the objects of study and the points of focus, as well as the research questions and methods employed to structure and answer them. One can appreciate in this example the sophistication and the complexity of accomplishing this task in a CALL environment.

Speaking more generally for CALL, an appeal to Vygotsky's work has been made to support the following techniques and approaches in language teaching and learning: cooperative or collaborative learning (Light, 1993; Warschauer, 1995a; Warschauer & Kern, 2000); teachers working with students on purposeful activity (Barson, 1997; Jones & Mercer, 1993); and learning in social groups and communities of practice (Debski, 1997). As McDonell (1992) observed, Vygotsky's theory supports a collaborative approach and cooperative learning, because it "analyses how we are embedded with one another in a social world" (p. 56), and because it is consistent with a view of teaching in which the process of mediation is central. In CALL, as elsewhere in education, there is evidence to suggest that Vygotskian sociocultural theory has been and continues to be highly influential.

Activity Theory

General Description: Education. Because *activity theory* is considered the contemporary formulation of Vygotsky's work, this section naturally follows from the last (Lantolf & Pavlenko, 2001). A number of scholars have contributed to the current view, including Engeström (1987, 1999), Leont'ev (1981), and Wertsch (1998). Engeström (1987) expanded on Leont'ev's model and defines the basic structure of an activity system in terms of six constituents—object, subject, tools, community, rules, and division of labor—operating on the three levels identified by Leont'ev (1978). For the purposes of this discussion, we concentrate on Leont'ev's analysis of activity; a useful summary was provided by Parks, Huot, Hamers, and H.-Lemonnier (2003):

> In developing this model, Engeström initially drew on [Leont'ev's] analysis of activity, defined in terms of three hierarchically related strata: Activity, Action, and Operation. At the uppermost level of this hierarchy, the targeted activity is viewed in terms of how the individual or *subject* (e.g., the teacher) initiating the action conceives of its underlying purpose or goal. At the second level, the targeted activity is further specified in terms of goal specific actions and the *tools* used to carry them out .... The third level pertains to the actual enactment of the activity (the *object*) in response to prevailing conditions. Whereas the first two levels are under the individual's conscious control, the operations, viewed as routinised procedures or strategies, function at the level of the unconscious. The actual *outcome* of the targeted activity is a function of how it is conceived by the individual and the tools selected to enact it. (p. 29)

Activity theory (AT) is a relatively complex and ambitious theory that aims to account for significant features of the learning context at different levels, from the level of the individual to the broader sociocultural context of teaching and learning. As such, it proposes a framework that identifies and locates the activity system as the basic unit of analysis—the activity system comprises a dynamic network of interacting and interdependent elements with its own cultural history.

Lim and Hang (2003) offered an excellent overview of the implications of this theoretical perspective for the use of ICT in education, especially in relation to the critical issue of successful ICT integration in schools. For Lim and Hang (2003) and for others, the attraction and power of activity theory as a framework is the theory's capacity to account for significant elements of the broader context of ICT use in education. Such contextual factors or variables lie beyond or outside the purview of other theories, so they are of little help in describing or explaining them. Lim and Hang (2003) stated:

> By linking the effective or ineffective integration of ICT in specific class-rooms with particular learning activities, situated within their larger sociocultural context, it is possible to build a detailed account of what the participants in the classrooms have done to make the activities successful, how the activities are supported by their larger sociocultural context, and what problems are encountered. (p. 50)

Thus, rather than a task, an interaction, or a student's direct experience of ICT, with activity theory the unit of analysis is much more broadly conceived in the activity system (Engeström, 1987; Leont'ev, 1981).

Activity theory is also considered powerful by its supporters because of its capacity to capture the dynamic nature of activity systems and changing points of focus over time—for example, occasions when there is a dislocation or breakdown of some kind—as well as salient features of the learning context that are considered to influence success but are not accessible via other theoretical accounts. Lim and Hang (2003) described how perceptions of ICT may change, from an *object* initially (when first installed), to become, over time, a *tool* that mediates teaching and learning once ICT is integrated into the school curriculum; however, if there is a problem of some kind with the technology, for the teacher or the student, the technology ceases to be a transparent tool and again becomes an object.

Tolmie and Boyle (2000) gave an example of the application of AT in the context of CMC use. In discussing what sustains a series of CMC interactions, they argued that it is not so much that users necessarily possess a good mutual understanding, especially at the outset, but instead that they are both motivated by a *shared purpose*. Activity theory provides a rationale for this idea by describing how, in an activity system, a subject (the individual) uses a tool (CMC) to pursue an object (a shared purpose). Furthermore, and crucially, AT says that subjects, tools, and objects carry their own history in the sense that the user's past experiences influence present perceptions of these artifacts. Therefore, on each occasion a new activity system is encountered or created, the user's knowledge is not invented anew, but rather is carried over or reconstructed, imperfectly, from past experience. Thus, it is this knowledge or shared purpose, rather than mutual understanding, that sustains the interaction, especially when disjunctions or problems occur. This idea of shared purpose and, as Tolmie and Boyle (2000) noted, the notion of cultures "stored up" over time in objects and tools, are parts of the contribution of AT to our understanding of ICT use in education.

Schulze (2003) appealed to AT for CALL in yet another way, in relation to human–computer interaction (HCI) and feedback. Schulze drew on the AT's analysis of activity, defined in terms of the three hierarchical levels of activity, action, and operation. This hierarchy provided Schulze with a mechanism for identifying significant qualitative differences between human–human and human–computer feedback and interaction. Schulze (2003) maintained:

> [C]omputers can theoretically carry out all operations, but never actions because they do not recognise aims, purpose, or position of the action. The differences between humans and machines can for our purposes (i.e., the theoretical description of human–computer interaction) be legitimately reduced to the distinction between actions and operations as is done in activity theory. (p. 446)

Essentially, actions are under the individual's conscious control, whereas operations—viewed as automatic procedures or strategies—are not, because they function at the level of the unconscious. This distinction captures a fundamental difference in the quality of the evaluation and feedback that a human and a computer are able to provide when they are interacting with students for the purposes of language learning. Whereas the human teacher will use discretion and weigh aim and purpose before giving a student feedback in class, the computer is programmed to provide an automatic response using a preset algorithm that operates automatically on the student input that it receives.

Blin (2004) drew on cultural-historical activity theory to provide an analytical framework to study the relationship between CALL and the development of learner autonomy. Learner autonomy was defined by Holec (1981) as the "ability to take charge of one's own learning" (p. 3). It is a complex, multidimensional idea that has both social and individual dimensions (Blin, 2004). Discussion of learner autonomy has maintained what might be described as a steady visibility in CALL (see Blin, 1999, 2004; Healey, 1999; Kaltenböck, 2001; Komori & Zimmerman, 2001; Murray, 1999a). It is only with the recent work of Blin that an integrated framework for the study of CALL and learner autonomy has been attempted, in this case informed by activity theory. There is insufficient space to do justice to Blin's analytical framework here, but a brief overview of the main ideas and the role of theory are given next.

In brief, Blin (2004) asserted that AT may assist in promulgating reform in education by making visible significant differences between "old" and "new" approaches, and helping to foster a transition toward a more constructivist, collaborative language-learning environment. She applied AT to analyze organizations undergoing change as a result of technology innovation. Specifically, Blin used Engeström's (1987) model of an activity system to provide an analytical framework and a terminology to describe language-learning activities as they unfold. By describing these activities as perceived by learners, using the terms and framework of AT (e.g., *actions, operations, objects, tools,* etc.), Blin was able to represent and describe the relationship between CALL and learner autonomy. The model assists in three ways: It provides a means to capture changing priorities during a language-learning activity; it allows the role and function of tools and procedures to be made visible in ways that other theoretical frameworks cannot, and it contributes to a better understanding of the nature of learner autonomy by showing how key variables might be operationalized and might impact a learner's autonomy under different conditions and at different times. In this way, the theoretical framework and the terminology enable the idea of autonomy to be made more concrete, so that in the future it may be analyzed, examined, and, potentially, measured objectively.

Finally, in summarizing the implications of AT for language learning, Lantolf and Pavlenko (2001) commented that it reaches far beyond a consideration of merely the acquisition of the forms of the language. The authors argued for a much broader agenda, and stipulated that language learning is "about developing, or failing to develop, new ways of mediating ourselves and our relationships to others and to ourselves" (p. 154). They concluded:

> In sum, from the perspective of activity theory, it is not necessarily the case that all of the people in language classes have the goal of learning the language and the reason for this is because they have different motives for being in class, because in turn they have different histories .... Cognitively, they are not all engaged in the same activity. And this is ultimately what matters, because it is the activity and significance that shape the individuals' orientation to learn or not. (p. 148)

Such a perspective emphasizes learner perceptions, and how learners interpret the activity that has been designed to involve and engage them. Activity theory, especially in its latest formulation, is still very new. We have seen in the few examples mentioned here that CALL practitioners and researchers are drawing on it in different ways, especially when they want to capture change in dynamic systems as a result of technology use and innovation. It remains to be seen how powerful and practical this theory will be for language learning.

Constructivism

"The Good, the Bad, and the Ugly: The Many Faces of Constructivism" is how Phillips (1995) entitled his perceptive review of this topic. He went on to liken constructivism to a "secular religion." Phillips used this analogy not only to emphasize its broad adoption as a large-scale movement and system of beliefs, but also to highlight its diversity and division, and to argue strongly that constructivism has many interpretations (or "sects") associated with it. These varying interpretations of constructivism can differ quite substantially one from the other, reflecting the broad spectrum of philosophical and theoretical positions that have at different times been used to inform it (see Jordan, 2004). There is some agreement in very broad terms about this general view of learning, which Phillips summarized by asserting, "[H]uman knowledge—whether it be the bodies of public knowledge known as the various disciplines, or the cognitive structures of individual knowers or learners—is *constructed*" (p. 5). But beyond that basic statement, interpretations tend to differ and follow rather different paths. Still, in his conclusion, Phillips was a little more specific when he noted that constructivism is "good" in the emphasis that "various constructivist sects place on the necessity for active participation by the learner, together with the recognition (by most of them) of the social nature of learning" (p. 11).

Dalgarno (2001) provided a useful interpretation of these ideas in his paper called "Interpretations of Constructivism and Consequences for Computer Assisted Learning." He defined the constructivist view of learning in terms of three broad principles:

- Each person forms their own representation of knowledge.
- People learn through active exploration.
- Learning occurs within a social context, and interaction between learners and their peers is a necessary part of the learning process. (p. 184)

For Dalgarno, these "basic tenets" of constructivism are generally agreed, but the implications for teaching and learning are not as clear. He went on to suggest three different interpretations of constructivism for computer-assisted learning (CAL) that focus, in turn, on:

- Learner-directed discovery using hypermedia, simulations, and microworlds to encourage active exploration within a virtual environment.

- Direct instruction while still allowing students to actively construct their own knowledge using guided hypermedia, cognitive tools (e.g., concept-mapping tools), and tutorial systems.
- Social interaction in the learner's knowledge construction process (with peers and teachers) and the use of computer-supported collaborative learning (CSCL) tools.

These understandings and widely differing interpretations of constructivism have carried over into the CALL area (Beatty, 2003; Felix, 2002; Rüschoff & Ritter, 2001; Shin & Wastell, 2001). In a special issue of the *TESOL Journal* entitled "Constructing Meaning With Computers," the editors wrote of cognitive and social constructivism (Healey & Klinghammer, 2002). In the discussion, they also emphasized the centrality of the learner in the learning process and the importance of the teacher in creating motivating authentic activities that involve investigation, discussion, collaboration, and negotiation. In the articles that comprise the special issue, each author drew rather differently on the constructivist idea, often listing overlapping sets of principles that underpinned the constructivist CALL learning environments they were creating. But, again, most emphasized the centrality of the learner actively constructing knowledge, sometimes alone, but frequently through collaborative tasks, using the technology to assist task completion. Thus, the constructivist view of learning contributes to the ways in which these authors understood what constitutes learning, how it is best achieved, and guidelines concerning the roles of the teacher and the learner and the technology in the process. In another interpretation of constructivism in the CALL context, Shin and Wastell (2001) commented on the concept in this way: "The essence of constructivism is to motivate learning by leading students to experience the individual and subjective satisfaction inherent in solving a problem that is seen and chosen as one's own. It is this view of learning that provides the general educational design philosophy of that methodological framework" (p. 519). In contrast, Vannatta and Beyerbach (2000) appealed to constructivism to provide a "vision of technology integration" for the use of technology in education where "an assortment of technologies and applications are used to enhance the creation of products, facilitate problem-solving, and assist exploration" (p. 134). As with other theoretical frameworks we have discussed, practitioners appeal to theoretical accounts in different ways for different purposes.

Multiple Theories and "Rare" Theories

This description of theory and CALL would be incomplete and give a false impression if it stopped here. We also need to include authors who draw on a number of theories simultaneously to motivate their work, and examples of authors who have used lesser-known theoretical perspectives in CALL. Obviously, the examples need to be limited here. Some multiperspectivist examples in CALL are shown in Table 5.1. Collectively, these examples illustrate the many different ways that theory is utilized in CALL. Clearly, we need to recognize that theory is used not only to motivate research, but also to guide design at many levels, including programs, Web sites, tasks, and various other kinds of learning environments.

It is interesting to note the recency of these articles; we were not able to locate very many examples of this phenomenon before 2000. Finally, to close this section, Table 5.2 shows some examples of authors who have employed theoretical perspectives for CALL that cannot be considered mainstream. Of course, Bloom's taxonomy is very well known in education generally, but its use has been surprisingly infrequent in CALL.

TABLE 5.1

CALL Practitioners Using Two or More Theoretical Sources in Their Work

Author(s)	Theoretical Source or Influence	Contribution of Theories (to provide for)
Gutiérrez (2003)	• Sociocultural theory • Interactionist account • TBLT	• A principled approach to task design • Systematic investigation of the processes of collaborative activity in computer-mediated tasks
Saarenkunnas, Kuure, and Taalas (2003)	• Computer-supported learning (CSL) theory • Multimodality • Discourse as social action	• The polycontextual nature of computer-supported learning • A participant perspective in computer-supported collaborative learning
Shin and Wastell (2001)	• Constructivism • HCI	• A general educational design philosophy for the methodological framework • A user-centered approach
Tschirner (2001)	• Situated learning (constructivism) • Interactionist account • Intercultural competence • Lexical learning • Emotional learning	• A principled approach to the development of oral proficiency in a multimedia classroom environment
Van de Poel and Swanepoel (2003)	• Theories of second/foreign vocabulary acquisition • Cognitivist approach • Models of human information processing	• Design and evaluation of CALL applications for vocabulary acquisition

TABLE 5.2

CALL Practitioners Using Lesser-Known Theoretical Models
and Frameworks to Motivate Their Work

Author(s)	Theoretical Source or Influence	Contribution of Theory (to provide for)
Belz (2001)	• Social realism	• A framework for exploring social and institutional dimensions of telecollaboration
Egbert (2003)	• Flow theory (Csikszentmihalyi)	• The conditions under which classroom computer tasks generate flow experiences • A framework for conceptualizing and evaluating language-learning activities
Jones (2003)	• Generative theory of multimedia learning	• The conditions under which multimedia annotations support listening comprehension
Svensson (2003)	• Bloom's taxonomy	• A principled approach to the creation and design of streamed media, hypertext, and virtual worlds
Weininger and Shield (2001)	• A discourse analysis model of different communication types (Koch & Oesterreicher)	• A precise description of the degree of proximity or distance in synchronous, text-based CMC (MOO) discourse

DISCUSSION

Introduction

In this chapter, we have seen many examples from the literature of CALL practitioners appealing to different theories. A summary list of reasons why this is so, drawing from the examples described in the book so far, follows.

Overall (design, teaching, research):
• To justify or legitimize a particular approach.
• To recommend or inform priorities, goals, structure, and procedure.

Design:
• To provide a principled basis for the design and construction of CALL materials (e.g., online courses, Web sites, CDs, programs, tasks).
• To structure and inform the creation of individual elements in any of the preceding (e.g., feedback, media selection, interactional features, resource availability).

- To help develop a view of the learner, individually and collectively.

Teaching:
- To provide a methodological framework.
- To guide technology selection and use in the classroom.
- To inform aspects of classroom management and organizational issues.

Research:
- To direct researchers to the important questions to ask.
- To provide a source of hypotheses to be tested.
- To provide a means of identifying and labeling significant variables.
- To help determine what counts as (language) learning (see Mitchell & Myles, 2004, p. 221).
- To account for significant factors or events in learning (with regard to the learner, the learning, or the learning context).
- To help locate the significant variables.
- To provide concepts and constructs.
- To provide an explanatory framework for second-language learning.
- To provide research tools and methods.
- To explain relationships.

There are clearly overlaps in the preceding list, but the list is a long one and illustrates well the potential value of theory to the CALL designer and teacher, and especially for the researcher. The verbs at the beginning of each item are instructive: to guide, to help, to design, to direct, to provide, to organize, and to explain. Theory can also illuminate and inspire. Overall, it is very clear that theory really does provide a very rich resource. However, like much else in CALL today, theory is not without complications. There are essentially three problem areas associated with theory selection and use in CALL: the number of theories now available that relate in some way to language learning, the variable status and robustness of each theory, and the problem of establishing clearly the range of applicability of each theory. These and related issues are considered next.

Theory for Design

Introduction.   CALL designers tend to use theory less prescriptively and more eclectically compared to researchers. Generally, they are comfortable drawing on a number of theories at different times for different purposes, whereas researchers typically limit themselves to a single theoretical orientation. In design, for example, Boyle (1997) noted that *presentation* design has its own problem space, that it is quite different from *conceptual* design, and it requires its own theories and rationale. Different facets of a fully realized CALL program, software, online materials, or learning environment may potentially draw on a large number of theoretical bases that pertain to different facets of the software application; in this sense, the situation with regard to theory is not dissimilar to language teaching, in which different strategies, tasks, and activities can be justified under the banner of a number of theories.

Although on the one hand CALL designers are trying to pursue the goal of creating their Web site, CD, or online resource—something of an art form in itself in contemporary multimedia CALL—designers are also confronted with the immediate affordances and constraints of the technology and software development tools to hand, and the time, knowledge, and expertise (either individually or as a group) that they have at their disposal.

The tension between the perceived goal and the affordances and constraints of the technology is quite real, and the journey of creating new CALL materials is often very much the story of negotiating a path that successfully reconciles these considerations. Theory's role in this process may vary according to the purpose of the project; for example, whether the goal is to test a particular theoretically motivated design element, or whether it is to create something for everyday use with language learners (see Jacobson, 1994; Levy, 1999a). With the former, a single theory is often the focus, whereas with the latter, a number of theories may be used simultaneously. In general, however, the role of theory in design tends to vary between two extremes: At one end of the spectrum the design may be very much theory driven, and at the other end of the spectrum the role of theory may be subordinated to practical matters that arise during the design process. This opposition was usefully discussed by Richard Coyne (1997) in his work on the philosophy of design in a technological context. A brief overview is given here because it not only illuminates the relationship between theory and design for CALL designers, but it also sheds light on the ways in which theory might be viewed with regard to practice in language teaching. The core of the discussion revolves around two philosophical positions—the rational and the pragmatic.

Rationalism.   Rationalism is any philosophy that emphasizes the primacy of pure reason. It gives priority to reason over other means of knowledge acquisition and justification, such as that gained through experience and the senses (Blackburn, 1994). Coyne (1997) asserted that rationalism is a philosophical position we are all involved with as far as our understanding of technology and of design is concerned.

In the rationalist tradition, theory is the principal concern and is held to be superior to practice. Thus, the theoretical leads and directs the practical and, generally speaking, practical concerns are of a very limited interest within this orientation. Coyne (1997) provided a useful summary:

It is assumed that practical implementation eventually follows from sound theory. Rationalism does not promote concern about immediate practical outcome. In summary, rationalism supports the view that technology is a product of theoretical inquiry and is subservient to it. This is a further instance of the widely held view that technology follows scientific development—technology is applied science. (p. 27)

In contemporary society, rationalism is very much the dominant view. Winograd and Flores (1986) added:

> The rationalistic orientation pervades not only artificial intelligence and the rest of computer science, but also much of linguistics, management theory, and cognitive science ... rationalistic styles of discourse and thinking have determined the questions that have been asked and the theories, methodologies, and assumptions that have been adopted. (p. 16)

This philosophical position has been and still is very influential in CALL research and design practice. What constitutes "good" research is still often defined within these parameters. In CALL, the influence of rationalism has been keenly evident in the literature, most notably in arguments that favor theory-driven design practice and formulaic, linear approaches to design that insist on distinct, sequential stages during the design process (see Coyne, 1997; Levy, 1997, 1999a, 1999b).

Pragmatism.   Pragmatism is a philosophical school of thought that contrasts starkly with rationalism. For pragmatism, Coyne (1997) asserted:

> This orientation is primarily concerned with what works—actions and consequences. In other words it begins with an understanding of technologies in the human context—how the technology fits within the day-to-day practical activity of people. A concern with theory (in the sense of generalisations, rules and formulas) is therefore displaced by practice. (p. 36)

The relative status of theory and practice—which dominates, and which is held subordinate—characterizes one of the key differences between the rationalist and pragmatic traditions in philosophy. For the pragmatist, theory making is a kind of practice; in other words, the two are intertwined, or in the words of Hickman (1992), "[T]heory no longer had to do with final certainty but instead, as a working hypothesis, with the tentative and unresolved" (p. 99). Pragmatists are highly critical of the idea that practice is theory driven. In presenting the position of the pragmatists, Coyne (1997) maintained that "for pragmatism, design is a kind of 'reflection in action'—needs are commonly identified in retrospect or during development ... rather than at the outset" (p. 11). He continued:

> In some quarters, the language of practice is considered to lack the sophistication of "theory talk". So people revert readily to "theory talk" to justify

their practices. We pay lip service to the importance of theory, while our research or design betrays our commitment to practice. This quiet victory of pragmatism is evident in the "pragmatic turn" taken in design methods. (pp. 48–49)

In terms of design, then, instead of theory dictating design from the outset, which then proceeds in a linear fashion, for the pragmatist the design process is more a question of reflecting on goals and constraints iteratively in a cyclical design process.

Another important characteristic of the pragmatic orientation concerns the way in which the technology aspect is considered to be a kind of envelope that surrounds and permeates the ways in which we think about it. Coyne (1997) stipulated, "[T]echnological thinking 'enframes' us into particular ways of thinking" (p. 9), and "[W]e are shaped by our technologies as much as we fashion them" (p. 7). In contrast with the rationalists, then—for whom technology follows science and is seen as an application of a theory or preset plan—for the pragmatists, technology permeates our thinking and our design conceptualizations from the outset.

For CALL, Goodfellow (1999) reflected a pragmatic orientation when he described a number of reservations about using a theory or a psycholinguistic model as a point of departure for design. He noted, "[T]he designer has to make a choice between theoretical validity and practical effectiveness" (p. 118). He continued, "Whilst a design may be based on psychological principles that are *a priori*, its educational effectiveness is often a more empirical matter, decidable only through an iterative process of development and evaluation, with attention being paid at all times to the context of learning and the learner's experience of it" (p. 118). This is a point that is made again and again in the CALL literature when designing for real educational settings, notably in many of the contributions to the *CALICO* special issue on courseware design (Jones, 1999b), and also in the design and development work of Holland, Kaplan, and Sams (1995), Hémard (1999), and Arneil and Holmes (1999).

A pragmatic orientation is also evident when the idea of metaphor is employed as a design technique; Coyne (1997) argued that metaphor as a focus in design discussion belongs very much in the pragmatic domain. Some metaphors that have been used for CALL projects include:

A personal journey (Brussino, 1999)
An arena (Mugane, 1999; Svensson, 2003)
A murder mystery (Nelson & Oliver, 1999)
A workroom, bookcase (Shield & Weininger, 1999)
A toolbox (Vanparys & Baten, 1999)
A conversational partner (Price, McCalla, & Bunt, 1999)

A bag or pack (Chun & Plass, 2000)
A conference (Debski, 2000)
A community (Felix, 2003b)
Dave's ESL Café (http://www.eslcafe.com/)
The VCU Trail Guide (http://www.fln.vcu.edu/default)
The CALL Cookbook (http://www.owlnet.rice.edu/~ling417/)
The Grammar Safari
   (http://www.iei.uiuc.edu/student_grammarsafari.html)

Although the value of metaphor in design should not be overstated because it can be restrictive if not used wisely, some of the benefits of using metaphor in CALL design are evident in the help such a tactic provides for:

- CALL designers, teachers, and students: In conceptualizing a new program, Web site, or language-learning environment at the outset.
- Students: In imagining what the language-learning experience might be like before they actually engage with it (e.g., goal, level of formality).
- Students: In appreciating the role of the online course component in relation to offline or classwork element.
- Students: In understanding the role and pervasiveness of the culture in the language and for raising cultural awareness.

Using metaphor in design also illustrates very powerfully that when designing for real educational settings, the designer is creating learning *environments*, not decontextualized learning *interactions*. The design and evaluation of the virtual learning environment, *Lyceum* (described in chaps. 2 and 3), is an excellent example.

Theory for Teaching

Language teachers, like CALL designers, have what might be described as a "difficult" relationship with theory. The nature of this relationship was described perceptively by Carter (1993), who from the practitioner's perspective described the negative connotations often associated with theory:

- Theory is usually opposed to practice, the former being seen as dry, intellectual and mental, whilst the latter is warm, rich, humane and fun;
- Theory is regarded as having a parasitical relationship to practice, which is the 'real thing', whilst theory is the crippling codification of something that is essentially anarchic, disparate and fluid;

- Practice is life, theory comes after life. (p. 31)

As Carter went on to explain, however, most teachers work with a "theory" of some kind, even if the teacher is not consciously aware of the detail or its precise relation to practice. Its impact is felt, nevertheless; for example, in the language that is chosen as a focus in the classroom, in the way the teacher handles feedback to students, or in the structure and content of the tasks and activities that are employed. However, like the designer, the language teacher is, typically, a pragmatist and is principally concerned with "something that works." Prior experiences and principles of best practice are very important. Still, for many teachers—especially those who are more reflective—knowledge derived from theory tends almost always to be judged against knowledge derived from practice.

When theory is used for teaching and CALL, it is generally better used as a guide, not as a prescription. And rather than employing one theory exclusively, language teachers tend to draw on a number of theories simultaneously: There is a distinct difference between the way in which theory is used in teaching and design compared to the single theoretical framework of many research studies. This approach to the nature, use, and application of theory for teaching and CALL was examined usefully by Doughty and Long (2003) in their discussion of task-based language teaching (TBLT).

Doughty and Long (2003) described TBLT as "an embryonic theory of language teaching, not a theory of SLA" (p. 51). They continued:

> And whereas theories generally strive for parsimony, among other qualities—to identify what is *necessary* and *sufficient* to explain something—a theory of language teaching seeks to capture all those components, plus whatever else can be done to make language teaching *efficient*. Language education is a social service, after all, and providers and consumers alike are concerned with such bread-and-butter issues as rate of learning, not with what may or may not eventually be achieved through a minimalist approach motivated exclusively by theory of SLA. (p. 51)

Thus, TBLT, as a theory of teaching rather than of language acquisition, is informed by a number of theoretical sources that blend into one another. In their paper, Doughty and Long (2003) used theory to derive 10 methodological principles, or "language teaching universals," for TBLT. These, in turn, are converted to pedagogical procedures, according to contextual factors determined by the teacher, the learners, and the learning context. Thus, the role of theory here is to provide a principled foundation, inasmuch as current research findings are able, for the methodological principles. These then are translated into pedagogical procedures according to the requirements of the particular setting in which language learning occurs (see also González-Lloret, 2003).

A good example of a more broadly defined set of guidelines that are drawn from a number of theories rather than a single one was presented by Egbert, Chao and Hanson-Smith (1999). They identified eight optimal *conditions* for CALL and used these conditions to organize the content of their book. The eight conditions are:

1. Learners have opportunities to interact and negotiate meaning.
2. Learners interact in the target language with an authentic audience.
3. Learners are involved in authentic tasks.
4. Learners are exposed to and encouraged to produce varied and creative language.
5. Learners have enough time and feedback.
6. Learners are guided to attend mindfully to the learning process.
7. Learners work in an atmosphere with an ideal stress/anxiety level.
8. Learner autonomy is supported.

Egbert et al. drew these eight conditions from a number of theoretical accounts and research that the authors argued are "the most widely researched and supported in the literature and make up a general model of optimal environmental conditions" (Egbert et al., 1999, p. 3). This theoretical diversity stands in contrast to the seven *hypotheses* that were derived directly from the interaction account, described by Chapelle. Compare Egbert et al.'s conditions with Chapelle's hypotheses. Chapelle (1998) presented the following hypotheses derived from the IA for developing multimedia CALL:

1. The linguistic characteristics of target language input need to be made salient.
2. Learners should receive help in comprehending semantic and syntactic aspects of linguistic input.
3. Learners need to have opportunities to produce target language output.
4. Learners need to notice errors in their own output.
5. Learners need to correct their linguistic output.
6. Learners need to engage in target language interaction whose structure can be modified for negotiation of meaning.
7. Learners should engage in L2 tasks designed to maximize opportunities for good interaction. (pp. 23–25)

In these hypotheses, the IA was used to provide a set of explicit assumptions for CALL research and practice. This list of hypotheses is narrower and more tightly focused on language interaction; this is not surprising, because it is derived from a more contained and clearly circumscribed theoretical position. In contrast, Egbert et al. (1999) drew on a number of theoretical perspectives, which they used to support and justify a set of conditions to guide practice. Note also the differences also in terminology: *hypotheses* (Chapelle, 1998), *principles* (Chapelle, 1999a), and *conditions* (Egbert et al., 1999). The label *condition* is of course stronger than *hypothesis*. It is indeed noteworthy that the more tentative *hypothesis* that appears in the research literature transforms into the more certain and unequivocal *condition* in teaching. This is precisely where the potential danger lies, as tentative research findings are translated into recommendations for effective language teaching. The line between using theory as a guide and using theory as a prescription is easily crossed. We feel that the application of theory should not be allowed to reach beyond the context within which it has been proved valid, and, generally speaking, theories of language learning are not sufficiently well proven and robust to be applied as *conditions* in language teaching, although that is not to say that their use in the form of guidance is not valuable and useful.

Egbert and Hanson-Smith (1999) structured their edited collection of 28 chapters on CALL research and practice around these eight conditions for optimal language learning. These conditions are in large part those recommended by the IA, with their focus on comprehensible input, interaction, negotiation of meaning, and language production through the completion of authentic tasks. The sixth condition also makes reference to the importance of learner attention for language learning, a further prerequisite that may also be tracked directly to the IA. In the Egbert and Hanson-Smith (1999) book, the discussion departed from the IA and drew on other theoretical perspectives or principles derived from best practice, most especially in the chapters relating to classroom assessment (chap. 15); learning styles and motivation (chap. 18); classroom atmosphere, including discussion of Csikszentmihalyi's flow theory, but also drawing on best practice, including dealing with technophobia (chap. 21); and learner autonomy, for which conditions for creating settings for autonomous learning are described (chap. 24). In all, a wide range of theories are involved in providing support for the eight conditions for an optimal CALL environment.

This theoretical pluralism is evident in CALL more generally, too. There appears to be a trend in recent CALL articles to invoke a number of theoretical perspectives simultaneously, especially when concerned with developing a conceptual framework for online teaching and learning. Perhaps multiple theoretical perspectives are an acknowledgment that no single theory is preeminent in describing the processes of language learning; alternatively, it may indicate that no single theory is sufficiently powerful to provide a broad and principled set of guidelines for the many decisions that need to be made in creating online teaching and learning tools.

A useful example was provided in Hampel's work (Hampel, 2003; Hampel & Hauck, 2004) in relation to audiographic conferencing in CALL. This example could have been chosen in reference to theory use in design, or indeed theory use in language teaching, as it is used here. Hampel (2003) culled from a number of theoretical bases in describing a conceptual framework for an online learning environment involving audiographic online tuition, as shown in Table 5.3. As Hampel introduced each theory, she described the way in which it contributed to her theoretical framework, as illustrated in the third column of the table. Clearly a multimodal, multifaceted online learning environment requires many interrelated decisions to be made, and each theory or theoretical framework makes its contribution in a different but complementary way. What Hampel presented, in essence, is the basis of contemporary theory for teaching online, which, it must be said, is a not inconsiderable accomplishment.

TABLE 5.3

Noted Theoretical Sources for the *Lyceum* Distance Language-Learning Environment

Theoretical Source or Influence	*Associated Names*	*Relevance of Theory to Project (to provide for)*
SLA—the interaction account	Gass and Varonis (1994), Krashen (1985), Swain (1985)	Interaction (tutor–student, student–student), negotiation of meaning, comprehensible input/output, focus on form, meaningful use of language
Sociocultural theory	Vygotsky (1978)	Collaborative learning, virtual communities, social computing, networked cultures, ZDP, mediation-using tools, meaningful tasks, real life relevance
Constructivism	Rüschoff and Ritter (2001)	A learner focus, knowledge construction (an active, creative, and socially interactive process), authentic tasks
Situated learning	Anderson, Reder, and Simon (1996), Halliday (1993)	Different language functions (ideational, interpersonal, textual)
Multimodality	Kress (2000), Kress and van Leeuwen (2001)	Multimodal communication (the use of several semiotic modes together—visual, verbal, acoustic), user choice to suit the task and different student learning styles

Theory for Research

When a theory is applied to research in CALL, it provides a context and a view of language and language learning. In providing a field of view, each theory prescribes a focal point and identifies factors that are believed to impact it. The point of focus and the factors—typically operationalized as variables—then become the centerpiece for any research study that derives from the theory. In this way, the theory drives and shapes the data collection, interpretation, and analysis. In other words, the theoretical account effectively identifies which elements are to be brought into the foreground and which are to remain in the background, or are to be disregarded altogether. Thus, in one theory of learning, for example, individual factors are paramount and social factors play a reduced role, whereas in another the situation is reversed and social aspects come to the fore. Each theory also defines its unit of analysis. In contemporary theory, units of analysis include the language-learning task, an activity, a project, or indeed the language learner.

Whereas CALL designers and language teachers tend to draw on two or more compatible theories in their work, researchers tend to use only one. This is usually because they want to test the theory in some way or use it for explanatory analysis. In the discussion that follows, we see how two different theoretical frameworks shaped the thinking of researchers as they investigated the language-learning potential of an online chat environment.

Earlier, when describing the interaction account and CALL, an example by Fernández-Garcia and Martínez-Arbelaiz (2002) was discussed; it was taken from the second issue of the *CALICO Journal* in 2002. Intriguingly, the article before this one in the same issue, by Mark Darhower (2002), also looked at the interactional features of synchronous CMC chat, but from a sociocultural perspective rather than an interactionist one. The two contrasting theoretical approaches illustrate well the choices that confront contemporary researchers when no single language-learning theory is preeminent and when more than one theoretical account lends itself to the job of description and explanation. Here, two approaches were taken that analyzed very similar kinds of data (chat) in very similar settings with students of a similar profile (native speakers of English learning Spanish), but that were motivated by two different theories. The theories made their presence felt most particularly in the vocabulary associated with the theory and in the selection and interpretation of the data. As a consequence, the readers' attention was drawn to different qualities and features in the data. Then, the interpretation and discussion proceeded against two frames of reference, each with their own set of priorities. In this case, even though the arguments were made in different ways and on different grounds, both studies produced results that supported the use of online chat for language learning.[4]

The two viewpoints demonstrate how different theories work to frame our understanding of human-human interaction for language learning and contrive to bring some elements into the foreground while leaving others in the background. They also determine the nature of the evidence required to show that learning has occurred. The different perspectives additionally bring with them different research methods and tools and, critically, they each have their own sets of technical terms. It is well worth considering these two studies in a little more detail and a good place to begin is with the language.

Table 5.4 lists the key terms used in each paper according to the theoretical orientation. It is striking how different the two lists are. The terminology in each list is almost entirely mutually exclusive. In other words key terms deriving from the theoretical orientation appearing in one paper do not appear in the other; the only exception perhaps is the word interaction. Quite literally the authors of the two papers are not speaking the same language.

Darhower's title described this paper as a sociocultural case study, and explored the "social interactive features" of online chat over a 9-week period with learners of Spanish (2002). The study used Vygotsky's theoretical framework and discourse analysis techniques to "describe and explain" the most salient features of the interactions that occur. Darhower applied the main tenets of sociocultural theory. Thus, learning occurred as a product of social interaction with a more knowledgeable partner (i.e., the teacher). Darhower also echoed Vygotsky's concepts of mediation and the use of tools, including the use of language as a tool. Working within this view, mediation was the "instrument of cognitive change" (Donato & McCormick, 1994). Darhower noted, "In the L2 learning context, the provision of positive and negative linguistic evidence by more knowledgeable peers and the development of learning and communication strategies are some of the mediational means by which the lower linguistic processes develop into higher forms of language use (i.e., discourse competence)" (p. 252). Darhower also spent some time discussing the concept of *intersubjectivity*— the building of a shared communicative perspective or context that participants gain through collaborative discourse. He described intersubjectivity as relating to the Vygotskian view of cognitive development as a byproduct of collaborative discourse. Darhower (2002) commented, "The establish-

---

[4]Norris and Ortega (2003) gave an example of a study where what counted as learning was dependent on the theoretical premises of the research framework to the extent that the same data could be interpreted as evidence that was either for or against acquisition, according to the theoretical approach taken. They concluded, "To summarise, what counts as acquisition (theoretically defined) … may be disputed by researchers from different paradigms" (p. 728).

TABLE 5.4

Two Theories and Their Vocabularies

Darhower (2002)	Fernández-Garcia & Martínez-Arbelaiz (2002)
• Social interaction	• Negotiation of meaning
• Sociocultural framework (Theory—Vygotsky, 1978)	• Discourse structure (Model—Varonis & Gass, 1985)
• Mediation	• Negotiation routine
• Interactional features	• Interactional modifications
• Intersubjectivity	• Input
• Collaborative discourse	• Modified output
• Shared context	• Exchanging (ideas)
• Identity exploration	• Response
• Community	• Interlanguage
• Social cohesiveness	• Optimal conditions for SLA
• Autonomy	• Characterise linguistic means
• Higher/lower mental functions	• Message comprehensibility
• Transformations	• Discourse generated
• Communicative behaviors	• Communication breakdown

ment and negotiation of intersubjectivity perpetuates collaborative discourse, which, according to the sociocultural view, is important for language development" (p. 249).

Pinning down exactly how learning is conceived (or "what counts as learning"—Mitchell & Myles, 2004, p. 221) is critical in assessing any theoretical framework. In this study, Darhower (2002) examined specific interactional features of chat, namely intersubjectivity, off-task discussion, social cohesiveness (including greetings and leave takings), identity exploration and role play, use of humor and sarcasm/insults, and the use of the L1 (English). Darhower attended most closely to seeking evidence of intersubjectivity and social cohesiveness. Through the presence or absence (plus measures of degree) of these two interactional features, he aimed to determine the quality of an interaction and, indirectly, the learning that occurred through it. In his conclusion, he emphasized the value of chat in enabling a dynamic learner-centered discourse community to be constructed. Following the sociocultural theoretical framework, with its emphasis on the social aspect of language learning, Darhower pointed to the importance of this learning environment for cementing relationships between learners and in developing each individual's sociolinguistic competence. This social view of learning differs considerably from the view of learning by Fernández-Garcia and Martínez-Arbelaiz (2002), described earlier.

Both studies used theory to support their rationale and to justify their work. Also similarly, they both drew on theory to identify desired features in the chat room interaction, and they used the methods of discourse analysis to gather the evidence from transcriptions of the interactions. However, in studying the transcripts, the two theoretical bases led the researchers in different directions. Whereas Darhower (2002) was looking for evidence of intersubjectivity and social cohesiveness, which were hypothesized to be important for language development and learning, and the development of social linguistic competence, Fernández-Garcia and Martínez-Arbelaiz (2002) were looking for evidence of interactional modifications and modified learner output. The differences are conceptual and lie in how learning is characterized by the theory and, on that basis, what data are required as evidence that learning has taken place. Whereas Fernández-Garcia and Martínez-Arbelaiz were seeking evidence of developing grammatical competence, Darhower in contrast was focusing on capturing evidence of developing sociocultural competence—two very different perspectives on language learning.

In contrasting the two studies, we find the focuses are different, and the perceived "learnings" or broader advantages as a result of the online experience are different, because the theoretical foundations selected ultimately took the researchers along different paths. Happily, both sets of findings are generally positive as far as the use of online chat is concerned. Setting one study against the other shows that a case could be made for the use of online chat in language learning from two different theoretical standpoints. Rather than being a problem, this might be regarded more appropriately as providing a stronger rationale for CMC. In fact, these two particular theoretical positions share important common ground in spite of their very different histories. Mitchell and Myles (2004) asserted, "Both interactionist and sociocultural research, in their different ways, show how the ongoing character of L2 interaction can systematically affect the learning opportunities it makes available, and have started to demonstrate how learners actually use these opportunities" (p. 258).

Still, the different theories available that may potentially be applied in CALL do pose a problem, especially for the language teacher or designer who is not an expert in the theoretical domain concerned and who—quite justifiably—may not judge knowledge of the intricacies of the theory to be a primary concern. The scope of the theories and the robustness of the research findings associated with them can also be problematic. Such issues have been discussed at some length with regard to SLA theory and research and its application. For example, Mitchell and Myles (2004), discussing the work of Ellis (1997), reported, "The findings of SLA research are not suffi-ciently secure, clear and uncontested, across broad enough domains, to provide straightforward prescriptive guidance for the teacher (nor, per-haps, will they ever be so)" (p. 261). Even for the subfield of Instructed SLA, they continued, "[I]nstructed SLA research is not identical with problem-solving and development in language pedagogy, and does not ensure a shared agenda between teachers and researchers" (p. 195). On the other hand, Mitchell and Myles added that we should certainly not forget the value of this association: "[P]resent SLA research offers a rich variety of con-cepts and descriptive accounts, which can help teachers interpret and make better sense of their own classroom experiences, and significantly broaden the range of pedagogic choices open to them" (p. 262). The same points, we believe, apply in CALL.

Ultimately, there is no substitute for balanced and perceptive theory evaluation. We concur with the advice of Ellis (1994) that we must "evaluate theories in relation to the context in which they were developed and the purpose(s) they are intended to serve" (p. 685). As he elaborated:

> An approach to evaluation that acknowledges that theories are contextually determined allows for an acceptance of complementarity without a commit-ment to absolute relativism, for it can still be argued that among theories constructed for the same purpose and context, one does a better job than an-other because it is more complete, fit the facts better, affords more interest-ing predictions, is more consistent with other theories, etc. (p. 685)

We have found publications that take a number of theories and weigh them against one another, providing an assessment of both strengths and weaknesses, to be comparatively rare. This is one reason why the very acces-sible discussion by Mitchell and Myles (2004) in their book on second-lan-guage-learning theories has proved helpful. In this book, at the end of every chapter, there are sections evaluating the strengths and limitations of the theory discussed in that chapter. Other useful works that can be recom-mended are by Long (1993), Ellis (1994, chap. 15), Gregg (2003), and Jordan (2004).

CONCLUSION

With rare exceptions, CALL designers and language teachers are predomi-nantly in the role of consumers as far as theory is concerned. For those in this group who see value in theory (and it must be said not all do), they re-view, select, and apply theories of language learning produced by others. This places them in a challenging and in some ways a vulnerable position, for three reasons.

The first overriding factor is that we live in an educational culture that tends to follow the rationalist tradition and values theory over practice. Whereas there is a strong case for this position when using theory for research (e.g., when testing hypotheses or the effectiveness of specific design elements), we believe the theory-practice nexus is more problematic when it comes to the application of theory or theories to language teaching or to design and CALL. Those working in the three areas of application—teaching, design, and research—call on theory from different backgrounds and seek to apply it in different ways. For the designer or teacher, the problems associated with using theory for design and language teaching can easily remain unseen or be overlooked. This can lead to oversimplification or misunderstanding, or more weight being given to a theory than is warranted. Whereas the researcher may be acutely aware of the shortcomings of theory, the CALL designer or language teacher may not.

The second challenge is simply one of choice, because of the number of theories potentially on offer. Following Mitchell and Myles (2004), we believe those entering the field of CALL today need to become conversant with a number of theoretical perspectives. Language learning is a complex process, and different theories focus on different parts of this process. In CALL design we have seen increasingly that multiple theoretical perspectives are used in order to accommodate and provide a principled approach for the design decisions that need to be taken. Designers draw on theories from human–computer interaction (HCI), multimedia, and presentation design, as well as from language learning or second-language acquisition. In multifaceted learning environments, it is not surprising that designers employ multiple theoretical perspectives, because design is a complex, multifaceted activity. In language teaching, too, where language learning is becoming more broadly conceived—aiming now not only to assist learners in acquiring the forms of the language, but also helping them to develop learning strategies, learner autonomy, and intercultural competence—the need for an appreciation of a wider range of theoretical accounts is clear. Having said that, it should be pointed out that there are dangers inherent in this "theory buffet" (although we don't believe this is true of *Lyceum*). With so many theories available, it is easy to find one or two that more or less fit CALL artifacts or plans that have already been completed, making theories into marketing tools rather than keeping them as guides in the design process (P. Hubbard, personal communication, May 2005).

The third potential obstacle is the danger of ideas poorly or partially communicated, as theory is translated into practice. There is ample room for the consumer to misunderstand or only partially understand any given theoretical perspective, and to overlook its shortcomings and the boundaries within which it may be applied effectively. The consumers' understandings of the research findings, interpretation, and argument that stand behind the theory is quite naturally not at the same level as those of the theory specialist, or of the researcher who has been directly involved in the application and testing of the theory, usually under very carefully specified and restricted experimental conditions. The consumers' priorities are different, quite understandably, yet the CALL designer or language teacher may remain in awe of theory because of our prevailing culture, which urges us to value theoretical knowledge over practical knowledge. There is always the danger of taking the theory as proven fact when that is simply not the case. The consumer or user of the theory may draw from its surface features only, or from certain ideas and not others, and accept uncritically what the theory has to say without really understanding why. Potential difficulties do not lie only with consumers, however; researchers also naturally tend to be advocates of the theories with which they work. Shortcomings are sometimes glossed over or not presented with a force that is due given the degree of certainty warranted by the findings. Generally, researchers are advocates and not critics of the theories they use. This is not intended as a criticism, but rather an acknowledgment that goals, priorities, and knowledge backgrounds vary among the designer, the language teacher, and the researcher. The priorities are different and the transfer, from theory source to consumer, is a delicate one. As writers such as Ellis (1994) and Doughty and Long (2003) have fully appreciated, it is a question of bridging the gap between theory and practice successfully by conducting and providing balanced theory evaluation and providing consumers with good advice—about where and when a theory can be confidently applied and where not, or where two or more theories may be used together advantageously.

Also, it is our firm belief that theory should not replace a principled approach derived from a designer's or language teacher's accumulated knowledge and experience gained through careful reflection on practice. Theory can, however, complement this knowledge and experience, it can make it more tangible and visible, and it can extend it by providing more precisely defined concepts developed through theory-driven research.

Theory and theory development must remain a goal in language learning and in CALL for all the good reasons that were listed earlier, at the beginning of the discussion section of this chapter. Whatever the current imperfections may be, they should certainly not deflect us from the goal of developing and working with resilient theories that are applicable to the context in view. However, we should also understand that the needs, backgrounds, and goals of the CALL teachers, designers, and researchers are very different, and this difference in perspective leads to a wide range of interpretations and uses of the prevailing theoretical models and frameworks.

Chapter 6

Research

This chapter seeks to build on the previous one while at the same time altering the angle of attack. Research in CALL is primarily conducted in relation to language teaching and learning (using technology in some form), and in relation to materials design. For language teaching, research includes gauging learner attitudes to and perceptions of different kinds of CALL, and theory-driven studies that aim to assess language learning in this context. Following from our discussion in the last chapter, theories range from those that focus more on the acquisition of the grammar of the language to those that involve acquisition of more broadly understood phenomena, such as intercultural competence. The shape and structure of the language-learning tasks involved in enabling learning to occur are often features in such studies. Research findings are frequently used to inform and refine pedagogical approaches too, especially in relation to synchronous and asynchronous CMC, as when students use e-mail and chat for language-learning purposes. In language teaching, CALL research may be circumscribed by classroom activity, classroom and computer laboratory work, or online tuition and distance learning. In materials design, the orientation and goals are a little different. In research design, the focus tends to be more narrowly defined and, as befits what is essentially a design discipline, the research often involves development. Studies frequently test design elements, not only to see if they have worked properly in a program or Web site (as described in chap. 3), but also to enable broader claims to be made about the most effective way to implement a design. The design of program help and feedback is a regular feature also, not only in general terms, but more specifically, when a researcher aims to customize help and feedback so that it better suits certain kinds of learners. Clearly, research in CALL covers a fairly broad area of work and interests. We are not able to cover all of them here, but we aim to provide examples of research strands in this field that are potentially productive. We adopt a particular approach in trying to accomplish this task.

Often, in books on research in language learning, the authors address their topic by describing research approaches or methods, one by one, chapter by chapter, with suitable representative examples in each case. For example, Johnson (1992) covered correlational, case study, survey, ethnographic, experimental, and multisite approaches to research. Nunan (1992) tackled the topic in a similar way, also featuring chapters on classroom observation, introspective methods, and interaction analysis; again, research study examples were included. Generally, this basic strategy for the discussion of research is logical and effective, especially given the complexity and diversity of the topic; it also follows a tradition in education and applied linguistics of introducing research methods in this way.

In this chapter, we take a different, although related, approach, one that we believe reflects more closely the current reality in CALL research studies over the last several years. Our general strategy is to allow actual examples of research to determine the shape, the terms, and the nature of the discussion; in other words, we invert the conventional approach and largely allow representative examples to determine the approaches and methods included, rather than the other way around. This is necessary because most CALL research does not fit neatly into conventional categories. Actually, it is difficult to find studies in the field that follow a single approach: Mixed-method approaches are common, especially studies that make use of quantitative and qualitative approaches. A variety of research tools are also used in many CALL projects, often in combination (see Murray, 1999b), and it makes sense to describe these tools in relation to the research projects in which they were used. In our case, then, a small number of research studies have been carefully chosen as exemplars to illustrate the research approaches and methods that have been taken. From there, inductively, we derive general principles for "useful" CALL research. Our approach to this chapter is consistent with that taken in earlier chapters of this book, in which examples of CALL activity in relation to the chapter focus were brought more into the foreground and used to drive much of the discussion.

It is clearly not possible to give examples of all the kinds of research in CALL, just as it was not possible in the last chapter to cover all the theories that have been used. Our survey is a limited one; research involving concordancing, for example, is not included. Nonetheless, we focus on six studies that we believe are representative and potentially useful strands of research. Three of the research studies are more oriented toward CALL in language teaching (O'Dowd, 2003; Taylor & Gitsaki, 2003; Tudini, 2003); the other three studies focus more on CALL design (Chun, 2001; Jones,

2003; Pujolà, 2001). Aside from survey research and comparative studies, which are considered separately,[1] the description and discussion parts of the chapter concentrate on the research undertaken in these six examples. The choice of these particular studies is, inevitably, a personal one. Even so, efforts have been made to construct a representative and instructive cross-section of some of the major strands of CALL research being undertaken, and the methods and techniques currently being employed. As a group, the studies represent large-scale and small-scale projects, quantitative and qualitative approaches (separately and combined), deductive and inductive methods, a range of modern data-collection procedures, and most important a good slice of research conducted in different CALL contexts, including chat, e-mail, multimedia language-learning environments, plus tutorial programs and generic applications used for language learning. In the description section of the chapter, each featured study is described briefly, and important characteristics of the approach, procedure, data, and tools are identified. Sometimes key results are included too, if they illustrate qualities of the research in an informative way. In the discussion section of the chapter, the qualities of the featured study are drawn out further, as being indicative of possible research strands and agendas for CALL. For a features matrix of the six main studies included in this chapter, see appendix B.

The studies chosen for this chapter represent work that we believe has the potential to contribute to research in the field in the longer term. In the opinion of the authors, the goals and approaches taken in these studies are judged to have sufficient quality, force, and potential to drive parallel research agendas; in many cases, they are already part of them. Clearly, the CALL community does not have the resources to pursue every avenue, so this narrowing down is important, especially given that so many study reports close with the admonishment that more research is needed. Clearly, comparisons and strategic choices do need to be made so that research work is productive in the longer term. Therefore, the six research studies discussed in this chapter—each with its topic focus, directions, and goals—are offered for consideration.

Note that the purpose of this chapter is not to provide a set of criteria for assessing different kinds of research. For this purpose, we highly recommend Johnson (1992), who compiled a concise set of evaluation criteria for a number of mainstream approaches to research, including correlational,

---

[1]Survey research and comparative studies are treated separately because of their frequent use in CALL research. Survey-oriented research is very common, especially for gauging learner attitudes and perceptions. Because some of these studies are of variable quality they deserve special attention. Comparative studies have also been evident in CALL from its beginnings, and have also attracted criticism, so these kinds of studies are considered as a special case as well.

case study, survey, ethnographic, experimental, and multisite studies. Each is presented with a detailed, worked example. By using this reference, or similar, alongside the description and discussion provided in this chapter, the goal is to provide the researcher with some practical guidelines and suggestions, and also perhaps some cautionary notes, when contemplating a research project in CALL.

DESCRIPTION

Survey Research

Surveys are frequently used in CALL research and, typically, they are employed to assess student attitudes and perceptions. This aspect may be broad or narrow in scope. The broader kind of survey attempts to gauge student responses to a mix of CALL activities viewed collectively. This collection of activities is usually referred to under some umbrella term, such as *web-enhanced language learning* (WELL) or *network-based language teaching* (NBLT). In contrast, the type of survey in which the scope is narrow tends to focus more specifically on student attitudes and perceptions in relation to a particular CALL product or artifact (e.g., program, CD, Web site, or learning environment), or toward an implementation of CMC-based CALL (e.g., a task, a mailing list discussion, or a collaborative e-mail exchange). Some examples of surveys used for research purposes reported are given in Table 6.1.[2]

Although a survey was featured at some point in all of the studies cited in Table 6.1, in all cases the survey comprised only one component of a broader data-collection process. This is a characteristic of survey use in much of the more recent CALL research, and is noteworthy for that reason. For example, Hoshi (2003) used interaction analysis and content analysis alongside a survey when examining a mailing list in an elementary Japanese language class. Kötter (2003) used a selection of transcripts drawn from log files of student interactions in a MOO-based exchange plus a combination of questionnaires at the end of the project. Winkler (2001) combined a think-aloud protocol with a questionnaire and a posttask interview when investigating how English learners used CD dictionaries. The combination of data-collection methods, sometimes referred to as a *network of methods*, allows for cross-checking to occur and has the potential to provide a more complete picture of the phenomenon in view. Typically, the survey component is used to monitor student perceptions and attitudes in relation to the matter at hand.

---

[2]Note that the items included in this table pertain to survey use for research; in contrast, Table 3.1 in this volume showed survey use for evaluation purposes.

TABLE 6.1

Goals of Research Studies Employing a Survey

Goal or Purpose of Research	Author(s)
To assess student attitudes and perceptions to CALL conceived more generally (e.g., toward learning on the Web, Web activities, Web-based classrooms, CALL, CMC, WELL	Ayres (2002), Felix (2001), Green and Youngs (2001), Sengupta (2001), Stepp-Greany (2002), Taylor and Gitsaki (2003)
To assess student attitudes and perceptions to CALL conceived more specifically (e.g., a task-based CALL activity, online tuition via audiographic conferencing, a mailing list for language learning, a task-based e-mail exchange)	González-Lloret (2003), Hampel (2003), Hoshi (2003), Kötter (2003), Leahy (2001), O'Dowd (2003)
To assess students' computer literacy and training needs	Barrette (2001)
To assess participants' perceptions of the usefulness of annotations and their experience of reading in a hypermedia environment	Ercetin (2003)
To investigate learners' attitudes and preferences toward different kinds of dictionaries (e.g., paper-based and CDs)	Winkler (2001)

A useful example of survey research was conducted by Taylor and Gitsaki (2003), who employed a series of surveys to assess student attitudes toward an English-language course for Japanese learners, referred to as a web-enhanced language learning (WELL) course. This study was carried out at Nagoya University of Foreign Studies in Japan. The students participating in the study were first-year university students, aged 18 to 19, majoring in social sciences, economics, arts, and design. They were at a pre-intermediate English proficiency level and studied English formally for 90 minutes each week. In their research report, Taylor and Gitsaki provided a description of the course, the training required, and the procedures followed. The data for the study were drawn from two questionnaires, distributed at the beginning and end of the semester, a difference of 14 weeks. The first questionnaire was used to elicit general information about the students' background and experience with computers; the second questionnaire required students to rate 27 statements relating to the use of the Web as a language-learning tool, using a 6-point Likert scale. The first set of results represented the findings of a survey administered in 2002 ($n = 112$). These results were then compared with two earlier surveys conducted in the years

2000 ($n = 117$) and 2001 ($n = 104$) with similar student cohorts who had completed the WELL course earlier.

Positive features of this study include the longitudinal research design, which was organized around a series of surveys conducted at intervals. A longitudinal approach to survey research is a strategy recommended by Krathwohl (1993), among others, as a way to strengthen the research design. Also, in this study it is noteworthy that language learning was conceived more broadly than merely as the learning of the language itself. Learning about technology and developing computer skills was considered to be important, too; in fact, in one reported result, the students' mean scores for learning computer skills via the Web was given a higher rating than was the mean score for learning English (Taylor & Gitsaki, 2003). Finally, on the grounds that the survey research has been repeated over a number of years involving a large number of students overall, the authors moved to make comparisons among their study participants, mainly teenagers, with teenagers learning English worldwide.

Comparative Studies

Although much research work has been directed at gauging learner attitudes and perceptions to CALL materials themselves (software, CDs, Web sites etc.), those working with technology in educational settings have also been drawn into undertaking broad, comparative research studies. This is a very different kind of study from the ones described in the previous section. Comparative studies have featured strongly in CALL, as they have done in educational technology more generally. Many of these research studies, from the early ones to the present day, have sought to prove unequivocally that the IT alternative is superior to teacher-fronted classroom approaches to language teaching that do not involve technology. Burston (2003a) referred to this phenomenon as the "burden of proof," whereby people using new technologies in education have been required to justify their use, or their expense, to those who fund the resources. Also, it seems that the demand to prove the effectiveness of new technology innovations far outweighs any requirement to prove the effectiveness of pedagogical innovations (see Chapelle, 2001). In other words, proof of the effectiveness of new technology in teaching and learning is demanded in ways that are not required of other educational innovations.

Typically, broadly based comparative studies of this kind are designed so that an experimental or treatment group, which makes use of computer technology in the lab, is compared with a control group, which is taught in the "traditional" way in the classroom; what "traditional" means exactly in this context is usually assumed or left unspecified (see Wilkins, 1972).

These studies have received much criticism. For example, in a detailed examination of the goals and nature of comparative studies, Pederson (1988) said, memorably, that comparative research should "forever be abandoned" (p. 125). More recently, commentators have referred to problems with comparative studies, although also trying to assess their potential value under certain conditions (see Allum, 2002; Levy, 2001).

One problem that has recurred repeatedly in comparative studies is the idea that CALL is a single "method" of instruction. Here, CALL is typically assumed to be a discrete and distinctive approach, usually of a tutorial kind, rather than a broad collection of approaches and techniques—a much closer approximation to the current reality in most cases. Rather than a single method, it is much more common to see a language teacher review and select an approach for a particular setting according to the resources available and the student needs (Chapelle, 2000). Broad-based comparative studies in the field are still evident, however. Examples include Chiao (1999), Levine, Ferenz, and Reves (1999), and Klassen and Milton (1999). Overall, there appears to be a very strong desire to compare CALL and the non-CALL contexts, and clearly it is a recurring feature of research in the field (see also Burston, 2003a).

Broad-based comparisons such as these, however, are not the only kind of comparative studies taking place in CALL; much more narrowly focused comparative studies are regularly conducted as well. Typically, comparative studies of this kind are aimed at assessing the effectiveness of specific software design features as their creators seek to reach a decision on the most effective design solution. Recent examples include Al-Seghayer (2001), who compared dynamic video and still picture presentation for aiding vocabulary acquisition. In this study, the performance of 30 participants was measured to judge the most effective mode of presentation under three conditions: printed text definition alone; printed text definition coupled with still pictures, and printed text definition coupled with video clips. A number of comparative studies have also focused on the use of different kinds of multimedia annotation (see Jones, 2003). Other examples include Hew and Ohki (2001), who studied the effectiveness and usefulness of animated graphical annotation (AGA) in a Japanese CALL program and compared the students' performance using the program with and without the AGA. Yoshii and Flaitz (2002) compared the effect of different annotation types on L2 incidental vocabulary retention in a multimedia reading setting. In this case, three types of instruments were used to assess vocabulary retention: picture recognition, word recognition, and definition supply tests. There are many examples of more narrowly conceived research studies of this kind. They constitute an important dimension of CALL research, as we see later in this chapter.

Researching Language Learning Through Chat at a Distance

The advent of the Internet and the use of CMC technologies for language learning have led to a steadily increasing volume of research in this area. CMC is a feature in many recent studies in CALL and one reason why it was identified as a dimension in chapter 4. In this section we look at a study involving chat (synchronous CMC), and in the next we examine a study using e-mail (asynchronous CMC). The study in focus here, conducted by Vincenza Tudini (2003) at the University of South Australia, investigated a group of intermediate learners of Italian interacting in pairs one to one with native speakers (NS) on a Web-based Italian chat program over two semesters. Nine intermediate learners of Italian (median age 19.5) and 45 native speakers (age not known) participated in the study. This study differs from others that look at the use of chat for language learning. Instead of setting up tasks using chat and learner–learner interactions in the classroom with the teacher, as is usually the case, Tudini looked at chat in terms of NNS–NS interactions, without the teacher present, in the context of distance learning. Chat potentially provides a solution to the problem of providing distance learners with oral interaction, if, of course, learners do in fact learn effectively in these circumstances. In this study, Tudini operationalized learning in terms of the interaction account (IA) described in the previous chapter. Therefore, in broad terms, if learners can effectively negotiate meaning and modify their interlanguage, as specified by the IA, then language acquisition is considered to have occurred.

In particular, the research explored the opportunities for negotiation of meaning in live chat with NSs, the principal triggers for negotiation and interlanguage modification, and whether public NS chat rooms that learners enter without teacher supervision are amenable to effective language learning. This study also brought to light interesting questions concerning the relationship between in-class and out-of-class work, as well as the degree of structure needed to provide effective language-learning tasks.

Tudini (2003) began by reviewing the IA literature and previous research on synchronous CMC, and included a discussion of the similarities and differences between mediated and face-to-face interaction. Drawing on Negretti's (1999) work in particular, she highlighted structures and patterns in the oral interaction, especially with regard to sequencing and timing, which are peculiar to the interlanguage produced in the CMC context. She also pointed to important differences between group chat sessions and the one-to-one interactions that are the focus in this study when 9 learners of Italian engaged with 49 native speakers.

An important dimension of this study, and one that could easily have affected the outcome, was the way the chat work was integrated into the course as a whole in terms of assessment (5% in Semester 1 increasing to

10% in Semester 2). Knowing that the work is going to be assessed is a form of pretask preparation that can affect the students' performance in the task (see Skehan, 1998). The information required for assessment was available in the form of logs of learners' interactions. This material was used by both teachers and students to provide "a snapshot of learners' interlanguage" (Tudini, 2003, p. 154) and as a basis for reflection for distance learners. A further point of interest is the way in which the assessment criteria were chosen so that they helped direct the students' attention toward meaning and mutual understanding over accuracy. Assessment criteria emphasized negotiation and modified output rather than accuracy, which, the author asserted, was more likely to encourage learners to involve themselves in negotiated interaction.

A corpus of data was created from the chat sessions and used as the basis of analysis for this study. From this, the author identified negotiation sequences, cases of modified output, and examples of recasts, in which the native speaker provided explicit feedback spontaneously in the form of a corrected version of the learner's original utterance. Tudini (2003) reported that negotiations were a feature of most NNS–NS chat sessions, although only 9% of the total turns taken were spent in negotiation sequences and adjustment of linguistic output.[3] Tudini explained that this figure was lower than learner-only chat studies and suggested that when interacting with native speakers, personality, educational background, disposition towards Italian learners, and gender made a significant impact. Thus, there is going to be variation from one native speaker to the next in a public chat environment. However, there is an authenticity evident when learners chat with native speakers that is important in itself, and may help learners make the transition from interacting with other learners to interacting in the wider world with native speakers. Tudini concluded the study with a discussion of further research directions, noting especially the need to identify foreigner talk in NS–NNS chat sessions, the need to further develop and refine suitable tasks for distance learners in the chat context, and the need to examine the issue of learner use of the logs of the chat sessions to see how much they may assist in helping learners identify or notice linguistic problems.

Researching Intercultural Learning Through a Collaborative E-mail Exchange

One research strand within CMC that has attracted interest is the use of e-mail for intercultural language learning. Projects that have been investigated range from short-term key-pal projects to much longer-term collabo-

---

[3]Tudini (2003) noted that learner-only chat studies (NNS–NNS) report negotiation sequences in around one third of turns, a much higher value than reported here (see also Pellettieri, 2000; Smith, 2003).

rative exchanges (e.g., O'Dowd, 2003; Warschauer, 1995a, 1995b, 2000a). The study described here is an example of a longer-term e-mail exchange involving five pairs of students located in Spain and the United Kingdom with a general language level described as "relatively advanced" (O'Dowd, 2003, p. 122). O'Dowd's (2003) study was basically a qualitative, longitudinal study that took place over a 1-year period. O'Dowd labeled it as essentially ethnographic (see also Warschauer, 2000a), but also reflecting the principles of action research (see Nunan, 1992). The data-collection techniques employed included participant observation, e-mail data, questionnaires, interviews, the researcher's reflective journal, and peer group feedback.

O'Dowd's main research question sought to identify the characteristics of an e-mail exchange that led to intercultural learning, in particular what leads some network exchanges to fail whereas others succeed. The study was very much embedded in the literature on intercultural learning, both offline and online. O'Dowd differentiated intercultural learning from intercultural competence, at least as far as this study was concerned, and concentrated on the former. Theoretically, the paper was positioned in relation to Byram's (1997) model of intercultural communicative competence. This model provided O'Dowd with a series of components that could be operationalized in the design of the tasks that were used in the e-mail exchange. Byram's model was referred to throughout the study as a means of determining whether intercultural learning had been achieved.

Ten kinds of task were used for the study. As might be expected with the intercultural learning focus, the tasks moved well beyond language per se; in fact, only one task—the "comparative expressions" task—overtly focused on language. The remaining tasks encouraged the learners to explore their own cultures and the cultures of their partners through activities such as discussion of their perceptions of the image of their countries abroad, particularities of their hometowns, experiences with members of the partners' culture, reflections on target-language texts, and comparison of reactions to a fictitious story.

Following the pattern of most ethnographic studies, rich data drawn from the study were presented to the reader in the form of extended e-mail extracts and the researcher's commentary. The researcher's commentary consisted of a description of the context of the particular e-mail exchange and, as relevant, a description of the particularities of the students involved. In this way, drawing on Byram's (1997) model, the researcher was able to identify successful and unsuccessful e-mail exchanges, describe in a principled way possible reasons for the success or failure, and for those relationships that worked, pinpoint evidence of intercultural learning. The research article closed with a discussion of the successful and unsuccessful e-mail exchanges. For those students whose experience was positive, the researcher listed the particular characteristics of the exchanges that enabled learners to successfully develop aspects of intercultural communicative competence. He also identified elements in the e-mails of students who were able to develop a successful and interculturally rich relationship, thus answering the research question.

Researching L2 Reading on the Web

One of the successes of CALL research thus far has been the result of concentrated attention on specific language areas and language skills. Table 2.1 in chapter 2 gave a listing of the skill or area with authors and example studies. With regard to language areas, the most sustained effort has been put into CALL and vocabulary learning; with the language skills, the most focused work has been on reading and writing, although more recently listening and to some extent speaking have increasingly played a role. In general, this research draws deeply from a literature that is built around the specific nature of the language area or skill, and the cognitive processes associated with how it is acquired or learnt. Typically, it involves researchers who have developed specializations in the area over an extended period of time. One such example is Chun (2001), who presented an article on L2 reading on the Web (see also Chun & Plass, 1997).

Chun's (2001) study was conducted at the University of California in the United States. It examined how 23 2nd-year German learners accessed information while completing reading tasks using a Web-based program called *netLearn*. The students employed a variety of online multimedia resources when completing the tasks, including a glossary, an online bilingual dictionary, and an audio narration of the text. After reading a text, each student wrote a summary that was scored according to the number of propositions recalled from the original text. Also, a tracking device recorded the students' actions and the resources encountered; in particular, it recorded the number of words looked up and the amount of time spent on the reading and writing tasks. A subset of four students did think-aloud protocols while they were reading online, and another group of seven students completed an interview after they had finished the online task; in the interview, the questions were designed to investigate the students' metacognitive reading strategies.

This study critically reviewed the substantial body of research on reading, offline and online, in considerable detail. In so doing, it moved from a review of the general findings on the reading process to research on reading in hypermedia environments, first more generally and then for L2 reading specifically. It also covered the research on individual differences in learning with hypermedia. The literature review indicated clearly what research had and had not been completed and theoretical and conceptual frameworks that underpinned the study were described. This background provided a clear foundation for a series of interrelated, explicitly stated research questions.

The study then thoroughly described the materials, participants, and the procedures, as one would expect in any professionally reported research study. Of special note here, in relation to CALL research specifically, is the reporting of aspects or factors specific to the context. In this respect, the materials section of the research report is of special interest and importance (often insufficient detail is given). In this study, the Web-based hypermedia program (*netLearn*) was described, as well as the operational capabilities and functionality of the tracking program employed (*ActionCatcher*). Furthermore, in any research in this field, the multimedia components are of special interest and should be described, as was the audio narration component in this study.

In contemporary CALL research, tracking programs of different kinds are being employed increasingly. Such programs provide for a further dimension of data collection that is very useful when combined with other datasets. In Chun's (2001) study, the tracking program, *ActionCatcher*, was custom built and written by an associate of the author. This is a possibility, if the expertise is available, but a number of commercial authoring systems offer their own tracking systems. Some more recent research studies in which tracking of some kind has been used include Pujolà (2002), Roed (2003), Hwu (2003), Glendinning and Howard (2003), and Levy and Kennedy (2004).

Briefly, Hwu (2003) looked at learners' behaviors in computer-based input activities elicited through tracking technologies, and provided a useful discussion on interpreting tracking data uses *WebCT's* tracking system, the system also used by Roed (2003). Glendinning and Howard (2003) described the use of *Lotus ScreenCam*, the product also used by Pujolà (2002), as an aid to investigating student writing. They said of its role: "*ScreenCam* captures information which logging cannot, for example, movements of the cursor. But other useful visual information, for example, pointing with the finger at the text on the screen, is not available in either technique" (p. 44). Finally, Levy and Kennedy (2004) used a custom-built system to capture student discussions via audioconferencing on conventional video; this system was used for both research and teaching purposes.

Researching Student Use of Feedback and Help

One important and developing strand of research in CALL concerns the investigation of patterns of learner behavior as learners engage with online tasks and tutoring programs. The focus is on what students actually do. Typically, in this strand of work, researchers use a range of tracking techniques in order to record learner behavior, as we saw in the discussion of Chun's (2001) study earlier. This may involve recording the resources that students use (when, in what way, and for how long); for example, tracking the students as they access online dictionaries or help facilities, and logging time spent and patterns of use. Increasingly, researchers are relating this data to the students' language proficiency with the goal of providing resources and guidance that meet the needs of the students at a particular level. A related area of interest, using similar tracking techniques, is recording a student's response to program-generated feedback and looking at how an initial incorrect response is revised before resubmission. The knowledge that this kind of research generates is used to inform the design and development of CALL programs and language-learning tasks.

Good examples of research being conducted along these lines were provided by Pujolà (2001, 2002) and Heift (2001, 2002). Pujolà looked at learners' use of and response to feedback (2001), and the language-learning strategies students used when they accessed help facilities in a Web-based multimedia program (2002). Similarly, Heift (2001) investigated learners' response to metalinguistic feedback, and considered error-correction strategies in a Web-based intelligent tutoring system (ITS). Then, in a second study using the same program, Heift (2002) analyzed the error-correction process again, this time in terms of learner control and its impact on the ways in which learners made revisions and accessed correct answers. As an example here, let us look at Pujolà's work in a little more detail, concentrating primarily on the 2002 paper, which focused on identifying learner strategies when accessing help facilities in a Web-based multimedia program.

In this study, Pujolà used a Web-based program called *ImPRESSions*. It was designed to help learners develop second-language reading and listening skills and language-learning strategies. Its design is embedded in contemporary theory and best principles of pedagogical practice; however, the design is not informed by one theory or perspective, but rather by a number working in unison. The program is learner centered and controlled, allowing students to move among four modules: newspaper, radio, television, and "experts" (the last item provides an explicit training component for students on reading and listening strategies). In this study, a prototype of the program was used with Spanish learners of English at the University of Barcelona in Spain. Specifically, 22 adult learners participated in the study over 6 weeks with approximately 100-minute sessions each week.

A central feature of this program is the design of the help facilities. These are divided into two main groups, which Pujolà labeled *assistance* (i.e., help for comprehension), including the dictionary, cultural notes, transcript, subtitles, play controls, and feedback; and *guidance* (i.e., help for tasks), which includes operational information and instructions and language-learner training—the component that houses the expert module that provides advice to the student. Pujolà's objective in this research study was to see how learners make use of the help facilities. More specifically, his goal was to observe the language-learning strategies that students used when accessing the different help options.

As with other research projects in CALL, Pujolà's data were collected from a number of sources including material gathered while using the *Lotus ScreenCam* software (which records computer screen movements), plus audio recordings and more conventional researcher observation. Posttask interviews were also conducted to capture data from students on perceived language-learning strategies, and questionnaire responses provided information on the learners' perceptions and attitudes toward the program.

With each help facility, Pujolà identified patterns of behavior, grouping them where possible. For example, the data revealed two main approaches when students accessed dictionaries to look up words. Similarly, two types of approach were identified when students accessed the cultural notes. Pujolà labeled these two groups "compulsive consulters" (those who consulted all the cultural notes) and "selectors" (those who had a more selective approach and seemed to be more aware of learning strategies and the structure of the text). Identifying patterns and labeling groups of students who seemed to behave in a similar way in a particular context was a feature of Pujolà's descriptive approach.

In his conclusion, Pujolà (2002) affirmed that the program catered for individual differences, including learning styles and strategies for independent language learning in the areas of reading and listening. He also argued that this research contributed in broader terms, beyond an evaluation study of the program itself, to CALL research and design in general. This is really what separates the design and reporting of this study from the evaluation studies of CALL programs discussed in chapter 3. Pujolà (2002) made broader claims about optimal design features that move beyond the confines of the particular program, *ImPRESSions*. He maintained that the conceptualization of help facilities into assistance and guidance is of general value for any implementation of CALL software. He also argued that the study was of importance more generally because it showed that such a program can be designed using a Web-scripting language. In this sense, the work may be considered a proof-of-concept design study. Pujolà demonstrated that an abstract design concept can be successfully translated into a working program.

Experimental Research: Choosing the Most Effective Annotation Design

In contrast to many of the other studies discussed in this chapter so far, this example by Jones (2003) is principally an experimental study that employs a control group and three treatment groups, and inferential statistics to determine the kind of multimedia annotations that best support listening comprehension and vocabulary acquisition. In line with many of the other examples highlighted in this chapter, however, Jones did not rely solely on quantitative, experimental techniques; rather, her research design also involved a qualitative dimension that used participant interviews to provide the students' "voice." This was used in combination with the experimental data to provide a richer and more complete interpretation of the phenomena in focus.

This study examined two hypotheses. The first hypothesis stipulated that students would recall more propositions from the text of a listening-comprehension text activity when they had a choice of visual and verbal (multimodal) annotations compared with students who completed the same listening tasks with single-type (unimodal) annotations (visual or verbal), or no annotations at all. The second hypothesis examined how students best acquired vocabulary under the different conditions and details of the variation from treatment to treatment. Accordingly, 171 English-speaking college students in second-semester beginner French at the University of Arkansas participated in this study. The students were required to listen to a 2-minute, 20-second historical account in French presented by a computer program. Participants were then randomly assigned to four listening treatment groups, with the aural text presented under different conditions:

1. No annotations for key vocabulary (i.e., the control group).
2. Visual annotations only (with an icon to view images).
3. Verbal annotations only (with an icon to view text-based translations).
4. Both visual and verbal annotations (both icons available).

Students had to complete a vocabulary pretest. On the following day, they completed the main experimental procedures by listening to the prescribed text, looking up all annotations available in their treatment, and summarizing all they could from the passage. They then had to complete a vocabulary posttest. Three weeks later, students were required to complete a further, unannounced, delayed vocabulary and comprehension text. Twenty students were also selected for interview; it was these results that formed the main focus in the research report by Jones (2003).

The quantitative results of the study supported both hypotheses. Students remembered word translations and recalled the passage best when they had access to visual and verbal annotations, reasonably well when they had access to one or the other, and poorest when no annotations were available. However, beneath this general finding there was much valuable detail documenting precisely the conditions under which the different kinds of annotations worked effectively. For example, even though the combination of visual and verbal annotations produced the best results, students also expressed, through the interviews, a desire for choice, and options for viewing material in both a visual mode and a verbal mode separately, according to the particular situation. Another interesting result was that many students believed that the verbal annotations (i.e., translations) were easier to process than were the images, because they required less mental effort to clarify meaning. However, those who accessed visual annotations outperformed those who did not, showing the value on this occasion of learners choosing the more difficult path for increased learning.

This kind of research is quite common in CALL. Also, quite frequently, annotation types are assessed in relation to language areas or skills; thus, Ercetin (2003) looked at the annotations students use while reading a hypermedia text, and Yoshii and Flaitz (2002) and Yeh and Wang (2003) assessed the effectiveness of different annotation types for vocabulary learning. The results of such research projects have the potential to feed directly into the design decisions of those in CALL-developing programs, Web sites, and online courses. This research also brings to light learner behavior, both collectively in describing broad patterns of use and effectiveness, and more individualistically in describing individual approaches and patterns of interaction.

DISCUSSION

Introduction

So far in this chapter we have described a number of approaches to research in CALL, and six exemplary research projects that aimed to display, as a group, a reasonably representative mix of approaches, methods, research tools, and procedures. This group of projects also displays collectively a number of characteristics that we believe all CALL researchers should aspire to in their work, or at the very least consider carefully. The research studies:

1. Are deeply embedded in the relevant literature, which is reviewed critically.
2. Include theoretical and conceptual frameworks to support the study.
3. Include research questions or hypotheses stated explicitly.
4. Describe the workings of any program(s) used (e.g., Chun, 2001—*netLearn*), complete with sample screens, when possible.
5. Focus on what students actually do, supported by a detailed description.
6. Employ a network of research tools to capture as completely as possible an understanding of the students' experience and learning.

7. Use a mix of quantitative and qualitative methods.
8. Include a detailed description of the participants and procedures.
9. Provide a detailed description of the materials used (e.g., Chun, 2001, in appendixes 1–5, included the reading texts, the propositions for the texts as scored by the two raters, sample excerpts from the think-aloud protocols, interview questions, and sample data from the tracker).
10. Include, when relevant, a tracking device to capture online activity and to complement other methods of data collection, in order to provide a more complete picture of the phenomena under investigation.

The chapter now continues with discussion of the research strands, with reference to the research projects selected as examples. In this section, the main goal is to widen the scope of the discussion, explore themes and issues pertinent to the research strand, and reflect on possible avenues for longer-term research. In addition, the strengths of the different approaches, methods, and procedures represented by the featured studies are considered, along with some of the drawbacks and possible dangers of adopting the particular approach. In this way, it is hoped that the reader will gain insight into the different research strands in ways that will be helpful when formulating research designs and in setting personal research goals and agendas.

Survey Research: Managing the Limitations

Survey research is one of the most frequent kinds of CALL research, so it is fitting to begin the discussion with this topic. Unfortunately, and it has to be said right at the start, the quality of survey research in this area varies greatly. Therefore, it is worth revisiting several of the common problems with this kind of research; some points are of a general nature, and some are particular to the CALL situation.

Many surveys suffer from problems of ambiguity or lack of detail. This problem may apply to lack of information about the object of the survey research, the participants, the materials, or the context. Problems arise with the object of the research when the purpose and scope is vague or unfocused. Typical examples include broadly based survey research involving technology in language learning. For example, Web-enhanced language learning, network-based classrooms, or CMC-oriented CALL, if left undifferentiated, are such broad and multidimensional concepts that any research relating to them is likely to be of little value unless specified further. It is like conducting a survey research study on classroom learning without further specifying what activities exactly are to be completed and researched in that environment. Web-based learning, for example, includes a vast assortment of activities, often within the same Web site, that may vary widely in value and effectiveness for any particular group of students, especially when integrated into a broader scheme of work. It is therefore meaningless to survey students on entire Web sites, like *Dave's ESL Café*, that have so many parts to them. For a research report to be meaningful, the separate activities, tasks, or CALL materials have to be described in some detail. Better still, the focus of the survey research needs to be narrowed down to a single activity or a small number of related activities. The same applies to the participants in the study. If we are not given any information on who they are and relevant background details (e.g., computer literacy, first language, target language level, etc.)—which we would really consider baseline information—it becomes almost impossible for the reader of the research report to draw any meaningful conclusions that parallel their own experience and that can usefully inform future research. Although the results may be handy for local evaluation purposes, they are not of sufficient quality or depth to count as research worthy of a wider readership. From the reader's point of view, in any research report, we are seeking results that are meaningful and reliable beyond the bounds of the reported study itself. The researcher can help here by narrowing the focus, being as objective as possible, supplying detail, and reporting on the limitations of the study as well as the strengths.

Another common problem, especially with CALL, is caused by *novelty effects*. This often occurs, for example, in circumstances in which a survey is distributed soon after the students have used the computer, technology, or a new software program or application for the first time, either in the classroom or the CALL lab. Note that this may apply to the hardware and/or the software; for instance, when a new mobile technology is used for the first time, or a new piece of multimedia software is first employed. As long as it involves first-time use of a new technology or application, there is the danger of novelty effects. Not surprisingly, the students often tend to react favorably to something new because it is a change and different from the norm; their responses to a survey given shortly after this new experience can easily reflect an overly positive image. However, the opposite can also happen if, for some reason, the new technology or application is not working properly or is problematic in some way for students. Novelty effects do not have to be positive, although they often are. However, of course, what we are seeking is a more reasoned and balanced view from the students on attitudes and perceptions. Usually it is better to distribute a questionnaire or conduct an interview after a reasonable period of use.

Learner training also makes a big difference (this point is discussed in some depth in the next chapter). Learner training in CALL helps prepare students for the long-term use of CALL materials and can have a positive effect on student attitudes and learning outcomes (see Hubbard, 2004b). If surveys are given to students after they have been properly trained in the use of the materials, and after they have had some time to try materials out, there is more likelihood that the responses to survey questions will more accurately reflect informed and appropriately balanced student opinion, and novelty effects will be greatly reduced.

In fact, many CALL researchers are alert to this problem, and to cross-check results they use surveys longitudinally—either twice (simply at the beginning and end of semester) or at regular intervals. Examples of researchers taking this approach include Barrette (2001), Leahy (2001), Sengupta (2001), Green and Youngs (2001), and Taylor and Gitsaki (2003). Sengupta (2001) gave an interesting example of how valuable repeated data collection procedures can be when investigating attitudes and perceptions on student work publicly displayed on the Web:

> Indeed the mid-course interviews and the end-of-course interviews with students indicated a major contradiction in student perceptions. The themes suggested that there was a depth of conflict felt by students in interacting on a public forum, while at the same time there were informed comments about what they had learnt from their peers in a way they could never have in the face-to-face mode. The conflict seemed to be directly related to the accountability that the public and archived nature of the network context essentially imposes on students—a responsibility that was not so clearly visible in the face-to-face context. All interviewees in the on-going interviews seemed to imply that much of the feeling of heavy workload was arising from the fact that the Web was making each individual much more accountable because all the evidence of participation was there on the Web classroom. (p. 120)

Other general dangers of survey research can be broken into roughly four categories. First, in the construction of the survey questions, there is the danger caused by determining in advance what is going to be relevant to the issue at hand: Other unpredicted but important phenomena can easily be overlooked. The inclusion of open-ended questions as well as closed questions can help here. Second, problems can arise as a product of the survey or of the question itself, as when a student misunderstands a question or an instruction but still goes ahead and provides a written response; with a questionnaire, there is no way that the researcher will know this has occurred simply by looking at the survey responses. Third, the students are often asked to rate some phenomena on a scale (e.g., their computer skills, the value of the Web, what they have learned, etc.). In practice, this is a very hard thing to do; also, some students are rather more modest than others. Fourth, another general problem is often caused simply by the students' eagerness to please the teacher, especially in circumstances in which the student is a newcomer to the country, the culture, or the institution in which he or she is studying. For all these reasons, for the reader of the research report, single-strike survey research results involving technology in language learning should be treated with caution. For the researcher, the limitations of survey research need to be acknowledged—as, in fact, they have been by many CALL researchers. Any weaknesses can be compensated for by repeating the data-collection process at appropriate intervals and combining the method of survey research with other methods of data collection.

Finally, student attitudes and perceptions of technology change through experience. Any survey conducted at one point in time can only provide a snapshot of attitudes and perceptions: It can give no indication of the stability or longevity of these views. In fact, we would say that views of technology in any setting are particularly sensitive to learner experience. Depending on the personality of the student involved and his or her general attitude toward the use of technology in language learning—from highly favorable to highly unfavorable—stable conclusions are always going to be hard to draw out with any confidence, especially if one is expecting some kind of consistency as new technologies and new software are encountered the first time and then used over a longer period. As a result, if one is going to conduct survey research, studies over a longer time scale are needed, preferably following the attitudes and perceptions of the same individual learner over time.

Comparative Research: Broadly Defined and Narrowly Focused

Comparative studies that aim to compare learning environments that each comprise many composite elements continue in educational technology and in CALL, even though the criticism leveled at these kinds of study has been considerable. After numerous studies that have compared an IT option with the "traditional" classroom, we now have a new incarnation of the broad-based comparative study in which the online, technology-mediated option is compared to the "face-to-face" alternative. Although the comments in this section are critical of such sweeping comparisons, the main point is to emphasize that there are *different kinds of comparative research* in CALL.

The most common kind of comparison is the one in which the experimental or treatment group uses the computer, the Web, or the technology in some way, while the control group has "regular" or "traditional" classes (e.g., Chiao, 1999; Klassen & Milton, 1999; Levine, Ferenz, & Reves, 1999). There are further designs that aim to distinguish between two kinds of media mode, for example, e-mail and chat (e.g., Sotillo, 2000), or ones that seek to identify differences between technology-mediated and face-to-face language-learning environments (e.g., Schultz, 2000). Sometimes sweeping comparisons are made; for example, Sotillo (2000) proclaimed, "[I]n the hands of professors who know what they are doing, online instruction is *superior* to face-to-face instruction" (p. 83, italics added). This kind of comparison was also evident in Blake (2000) and Kitade (2000). Blake (2000) offered a good example:

> Rather, this study assumed certain advantages common to all CMC over face-to-face oral exchanges that have already been reported in the literature, namely that CMC constitutes:
>
> (a) a text-based medium that amplifies students' attention to linguistic form
> (b) a stimulus for increased written L2 production
> (c) a less stressful environment for L2 practice
> (d) a more equitable and non-threatening forum for L2 discussions. (p. 123)

As mentioned earlier, we generally believe that broad-based comparisons should be avoided in evaluation studies. They are problematic because of:

- The assumption that the CALL, CMC, or the technology option constitutes a distinct method, and is not a composite of techniques and approaches.
- The assumption that "traditional" or "face-to-face" teaching and learning constitutes a distinct method, and is not a composite of techniques and approaches.
- The assumption that individual differences do not make a significant impact.
- Lack of information in such studies about the CALL programs or generic computer tools used; the organization, content, and activities in the classroom; and the language-learning tasks.
- Uncertainty over what is meant exactly by the "traditional," "teacher-fronted," or "face-to-face" classroom.
- Uncertainty over whether the teacher is present when the students are using the technology-oriented alternative, and, if present, the nature of the teacher's role.
- Novelty effects in short-term, cross-sectional studies.

Broadbrush comparisons are generally not helpful, because they tend to collapse or gloss over important factors for the sake of a simple statement that supposedly "proves" that X is better than Y. It is not surprising that many of these studies, when analyzed more closely, have inadequate operational definitions and procedures. On the other hand, and this point should be emphasized, more narrowly defined comparative studies remain very useful (see Levy, 2001).

In particular, we want to draw attention to a kind of relation that might best be regarded as a minimal, focused comparison between media options. We believe these can be helpful and effective because, in these examples, the comparison occurs within a clearly defined, shared context. In each instance, the only difference between the two conditions is a change to the one design element that is being tested; for example, the type of multimedia annotation that is available. In other words, we are not talking about broad comparisons between online and offline learning, or language teaching with CALL and language teaching without it. Instead, these comparisons are narrowly focused, and both conditions operate in a context that is demonstrably shared in all respects, except for the independent variable being tested. For example, the three conditions compared by Al-Seghayer (2001) all share the context of a hypermedia learning program, designed by the researcher for reading comprehension; the comparison, then, takes place within that context. Similarly, for Hew and Ohki (2001), the study was conducted within the shared context of the JCALL program.

It is not, then, the case that all comparative studies are ineffective and should be discarded altogether. There are different kinds of comparative study in CALL, from those that attempt to make broad comparisons—between online and offline learning, CALL and non-CALL contexts, or technology-mediated and face-to-face settings—to those that are much more narrowly focused, and in which the comparison is made within a shared context or setting. Whereas the first kind of comparison is generally to be avoided, the second kind of comparison is to be welcomed, because such comparisons facilitate a focus on specific design features and, by making the comparison, we can see which design feature works most effectively.

Theory, Research, and Pedagogy

One of the fascinating aspects of research is that sometimes the same data can be examined from different theoretical perspectives, often with interesting results. We saw this earlier, in chapter 5, looking at the theoretical dimension of CALL in the context of online chat data, when the contrast was made between the theoretical approaches and interpretations of Fernández-Garcia and Martínez-Arbelaiz (2002) and Darhower (2002). Sometimes description and analysis from two complementary theoretical perspectives may provide a deeper understanding of the phenomena in view.

In our earlier example for this section, Tudini (2003) used the interaction account (IA), or more specifically the interaction hypothesis, to highlight sequences hypothesized to be important for language acquisition. The same data could also have been described and analyzed according to conversation analysis and discourse analysis techniques. This was the approach taken by Negretti (1999), who investigated nonnative speakers' chat sessions in English. Negretti completed her analysis of the interactions from a conversation analysis perspective. She was interested in the overall structure of the interaction and sequence organization, including openings and closings, turn taking, and expression of paralinguistic features and some pragmatic variables as well. With the data Tudini reported, one could use a Vygotskian perspective and look at sequences in which the native speaker provisionally took on the role of the teacher by offering explicit correction or guidance of some kind toward some aspect of the language or culture. This idea of a dual perspective was noted in passing by Mitchell and Myles (2004) who gave a good example with respect to a NS–NNS interaction extract. They argued:

> From an input or interaction perspective, such passages would be inter-preted as instances of negotiation of meaning, conversational repair, etc., and would be seen as maximising the relevance of the available input for the learner's acquisitional stage. From a Vygotskian perspective, it would be ar-gued that we are witnessing *microgenesis* in the learner's second language sys-tem, through the appropriation of a new lexical items from the scaffolding talk of the native speaker. (p. 210)

If the researcher's desire were to bring an even broader context into play, then activity theory might be employed. This theoretical base would allow the researcher to focus on such aspects in CALL as curriculum integration, technology breakdown, and breakdown recovery, and aspects of learner autonomy.

As discussed in the last chapter, the researcher's theoretical position is critical in determining what factors are brought into the foreground and what factors are placed in the background, or left out altogether. In order to select a theoretical point of departure, a CALL researcher has to answer three important questions: What exactly is language learning considered to mean; what constitutes the unit of analysis, and, relatedly, where are the boundaries of the learning context to be set? If one takes language learning to be the learning of the language per se (e.g., the grammatical features of the language, often considered to be the core of language learning), then one might well be drawn to a theoretical orientation such as the IA. How-ever, if one sees language learning in a broader sense to include culture learning, for example, one will need to use a more appropriate or encom-passing theoretical framework (e.g., Byram's model for intercultural com-petence—see O'Dowd, 2003). Equally, if one is more concerned with the social aspects of language learning then this focus will lead to another theo-retical framework again. In our opinion, it is not so much a case of right and wrong, but rather a question of one's point of focus and the theory's power to describe and/or explain the phenomena in view. In some ways, contem-plating alternative viewpoints should be expected. The nature of language and language learning are complex phenomena, and in many ways we are only just beginning to make headway in understanding them.

Taking a more socially oriented perspective on Tudini's (2003) study, one quickly sees features in an interaction that arise when the two participants do not know one another. There is plenty of evidence in the data to indicate the efforts of the two participants as they seek to get to know one another, in choosing topics and developing themes, and in the way they respond to one another. For the participants, naturally enough, this is their overriding concern as their relationship develops. Hence, although there are examples of recasts and other kinds of corrective sequences given spontaneously by the native speaker to the language learner, it is not surprising that these instances are relatively infrequent, just 9% compared to much higher values in learner–learner chat interactions. The priority is on engagement, showing interest and responding authentically in the dialogue.

It is also evident that Tudini (2003) was seeking to use this research to inform her pedagogy in the chat context. She sought to identify and encourage the kind of interactions required for language learning, wishing to improve on the 9% reported in the previous paragraph. With this in mind, she noted, "In order to encourage more goal oriented, cross-cultural useful and lengthy discussions with NSs, a focus on intercultural issues has been introduced as an element of chat tasks for both internal and external students" (p. 154). She continued, "So in the second stage of this project, learners have been asked to quiz chat participants *more assertively* to investigate Italian chatters' views on various topics, including differences between Italian, Australian and Italo-Australian perceptions of family" (p. 154, italics added). Learners are also encouraged to reflect on the chat experience in order to become more aware of their own language-learning processes.

In this study, research was very much informing pedagogy. Tudini (2003) concluded that both NS and NNS partners are of value for language learners in chat interactions, although for slightly different reasons. She explained that open-ended tasks of the kind used in this study only seemed to elicit self-repairs in learner–learner interactions. To move beyond this, following earlier findings from the literature, Tudini argued that specially designed two-way tasks are required to initiate the kind of negotiation needed for language acquisition. This leads us to consider the different kinds of tasks needed in the two learning contexts: one more open and receptive to an interaction with an unknown native speaker who may have no interest in language learning, and whose interest needs to be engaged; the other more closed and focused to encourage learners with a much more limited set of language resources at their disposal to interact in more focused ways that perhaps enforce certain kinds of interaction maintained to facilitate language learning.

How can we benefit from further research in this area? Repetitions of the study would be helpful with larger groups and including NNS–NNS and NNS–NS combinations. Further studies could also help refine the kinds of tasks needed in the two settings. Also, in NNS–NS interactions, it would be helpful to locate occurrences of foreigner talk in the chat logs, as Tudini (2003) suggested. Looking at different chat rooms in Italian, and then a variety of native speaker chat rooms in other languages (including, say, Asian languages) would be beneficial, too. It would also be helpful to analyze the data from different theoretical perspectives and then perhaps to consider the results side by side in the hope of reaching a richer and broader sense of the benefits of this kind of interaction; it might also be possible to refine the theories as well. Finally, the issue of transfer arises: To what degree does competence acquired in the chat room transfer to face-to-face contexts? Here, it seems case studies would be helpful, using a network of methods to track a small number of students over a longer period of time.

The Online Environment: New Skills, New Roles, and a Broader View of Language Learning

O'Dowd (2003) described his research study as ethnographic. The ethnographic approach to research has as its central tenet "the belief that the context in which behaviour occurs has a significant influence on the behaviour" (Nunan, 1992, p. 53). In this case, the context of the study is that of an intercultural e-mail exchange. Here, as on many other occasions through this book, the characteristics of the context are held to make a difference; in other words, until proved otherwise, an intercultural exchange conducted via e-mail is considered to be different from one conducted entirely in a face-to-face setting.

It is not difficult to recognize that intercultural learning is likely to be a rather different experience in online contexts compared to face-to-face encounters. In online situations, typically, it is the language teachers who make the initial contact on behalf of the student groups who will ultimately be communicating with one another. Like O'Dowd, the language teachers involved may have met through an online mailing list that specializes in facilitating partnerships (the IECC Network—O'Dowd, 2003), or they may have met at a conference, for example. Then, the two teachers discuss the project and reach an agreement to set up the collaborative exchange between their respective groups of students. From the students' point of view, typically, there is no graduated, introductory process with their partners, as there would be in a face-to-face setting. When students from different cultural backgrounds meet in the same language classroom, a good teacher can orchestrate a structured approach to any culture-learning activities;

also, the students know these relationships are important and will last for the duration of the class.

When students begin a collaborative exchange via e-mail, they are very much working with a blank canvas. However careful a teacher's preparation, no social convention dictates how the one-to-one relationship between partners might proceed and evolve. Current research tends to suggest there are hits and misses. The personalities of the students are likely to come to the fore unless the task is very narrowly structured, and research indicates that students move away from the set task very quickly when not directly supervised. From the student's point of view, one can begin where one wishes, one can choose priorities, and once can represent oneself in whatever way one pleases. Essentially, the students involved express their identities. O'Dowd (2003) showed that sometimes the circumstances can lead to frustration when partners feel that their views are not being heard or understood; on other occasions, the exchange can work well and develop very effectively. Overall, the results are very difficult to predict at the outset.

Looking at the examples of cross-cultural e-mail interactions in O'Dowd's study, the reader can easily detect the students' use of the medium to provide an opportunity:

- To help their partners understand their language and culture.
- To give information about the native language and culture.
- To represent their own language and culture as well as their own individual views.
- To "fight" against stereotypes.
- To address perceptions of the image of their country and culture abroad.
- To explain perceived national characteristics.
- To convince their partners of the "rightness" of the viewpoint.
- To negotiate through perceived differences.

In the e-mail examples that O'Dowd (2003) presented, clearly the language was in the service of communicating, explaining, and sometimes defending a social and cultural identity. The defense of social and cultural identity is often overlooked in research reports, which sometimes tend toward a more neutral characterization of an intercultural e-mail exchange, one that conveys the idea of presenting, exploring, and expressing cultural and social identity, but not explicitly defending it. This brings to light power relationships between cultures in contact via the e-mail medium. Some of the examples from O'Dowd's set of e-mails demonstrated convincingly the force of the views expressed and the depth of feeling involved. Often, they were deeply integral to the student's sense of self and cultural identity.

The power of such research reports undoubtedly comes not from some kind of statistical reasoning, but rather from the vividness of the examples. Their authenticity and the student's intentions in writing the e-mails shine through. One advantage of this kind of study, which applies to both ethnographic and case study research, as Nunan (1992) observed, is that "it is 'strong in reality' and therefore likely to appeal to practitioners, who will be able to identify with the issues and concerns raised" (p. 78). The strength of this kind of research also derives from the researchers' rich description and understanding of the context, and their ability to convey this to the readers, who need to be convinced of its authenticity and potential value. If the logical relationship that the researcher is able to draw between the data interpretation and the conclusions is also convincing, the readers are provided with examples from which they can easily draw parallels for their own work in a similar context.

These examples demonstrate clearly the student's purpose and motivation for sending e-mails. Any language that may or may not be learned is incidental and subservient to this purpose. From the students' perspective, the goal is not language learning so much as to inform, explain, and convince their partners of a particular viewpoint concerning their social and cultural identities. The language is very much in the service of an individual's communicative goals. For the language teacher who believes that language learning should be about cross-cultural communication and understanding, the promulgation of language and cultural diversity, and cultural pluralism, where many authentic cultures flourish side by side, the raw data and interpretation provided in O'Dowd's (2003) study will be highly valued.

Much depends, however, on the ability of the students, and their teachers, to negotiate successfully through difference, and their openness to questioning their own views and attitudes. The examples in the O'Dowd study amply demonstrated that sometimes partners will be receptive and open and will work through differences, and sometimes the relationship will founder because of an inability to see the other's point of view. In such situations, the longitudinal nature of the research becomes very important. In this regard, we would be eager to follow students over a period of time to see if the various pedagogical techniques available—distancing and role playing, for example—could be employed to encourage and develop intercultural language learning and intercultural competence, especially for those students who find this kind of interaction very challenging at the outset.

It is clear that the students need guidance in how to initiate and sustain a relationship online. O'Dowd (2003) described the importance of the teacher's role in giving guidance to students during their e-mail exchange by helping them to learn interculturally from their partners and how to sustain e-mails if some kind of difficulty should be encountered. However, the teacher's role is also crucial even before the e-mail exchange begins. The students need to be guided on how to initiate and develop a relationship on-line (i.e., they need training). E-mail exchanges will not always work out. O'Dowd's study showed that individual differences and personalities exert a strong effect, sometimes negatively, if one partner will not listen and respond to the views of the other.

This study also demonstrated that students have their own personal goals in language learning. Whatever the objectives of the teacher and how-ever well the tasks may be designed, the students still have their own goals, their own views about the tasks they are assigned, and their own reasons for being in the language classroom. What O'Dowd's (2003) study illustrated sharply is that individual student needs have to be reconciled with social conventions in the online environment if a successful relationship is to de-velop, just as they do in the offline setting, in the language classroom and elsewhere.

Finally, from such research, readers learn about the nature of the tasks re-quired, and the need to keep learners on task when strong personalities are involved. Learner training in initiating and developing online relationships is also key (see Hubbard, 2004b). Further research has the potential to lead to more effective tasks—perhaps tasks that are more shaped to the particular needs of individuals. The research can clearly inform pedagogy as well.

Research Through the Language Areas and Skills

Often in the CALL literature one finds research and design projects in which the field of view centers on a language area or a macroskill (a point we mentioned earlier in chaps. 2 and 3). This modular approach to design and research has been derived from perceptions of the strengths of technology and the roles it might perform in language teaching; and also, perhaps, the influence of linguistics, which has traditionally viewed language as a num-ber of fairly discrete areas or levels. This idea is developed further in the next chapter.

Should language be partitioned for the purposes of teaching and learn-ing? This is a fundamental question in language teaching as a whole and in CALL. Some might argue that it should not be partitioned at all, and that an integrated and holistic view of language is essential at all times. Others em-phasize the need to at times focus on language in its parts, so that some-thing that is large and complex becomes more accessible and manageable for the purposes of learning. This approach may also allow the language teacher to provide learners with language and learning activities more at-tuned to their specific needs or goals (e.g., academic writing). Such an ap-proach has the added advantage that all students and teachers know what is meant by *grammar, vocabulary, pronunciation, listening, reading,* and the other skills. Definitions of language skills and areas tend to be far less ambiguous for language teachers and students when compared with alternative terms, such as *communicative competence* or *focus on form*, which are frequently used in the academic literature on language learning. This is an important practical argument for a focus on language areas and skills that is often overlooked, unfortunately. Thus, there is less of a conceptual difficulty in dealing with language in its parts. In fact, it is probably harder to know what language is in its totality than to understand what it is from the point of view of its constituent elements.

As far as research and design are concerned, once the decision has been made to focus on a language area or skill, the likelihood is that theories and principles relevant to this component, rather than a more general set, will be utilized. Thus, Goodfellow, for example, referred to "lexical CALL" and described six design principles for computer-aided vocabulary learning (1995). These principles were not drawn from work in second-language acquisition more generally, but rather from theoretical work and research in vocabulary learning in particular. Much focused research work has been conducted in relation to vocabulary learning, especially concerning the structure and operation of the mental lexicon. One assumption in a discrete element approach such as this, however, is that the subcomponent may be studied effectively in isolation, and then that the skills acquired may be subsequently recombined and successfully integrated into the larger context (see Wilkins, 1976).

Also, the capabilities of the technology itself are instrumental in determining which sub-components are likely to become the focal point. The Ng and Olivier survey (1987) attempted to identify the aspects of language that the computer can best address. In order, they listed spelling, punctuation, vocabulary, reading, writing, discourse, listening, sociolinguistic appropriateness, and pronunciation. Since this early survey, increased attention has been given to listening and pronunciation (see Holland, 1999). What becomes clear from such listings is that the computer is more suited to handling some elements than others.

The CALL environment, as it stands, continues to provide well for receptive skills, reading, and listening, and for the manipulation of the written word through the word processor and e-mail, for example. Text and voice chat provide speaking or oral practice. Vocabulary is also well catered for. In fact, Table 2.1 in chapter 2 demonstrated that, to a greater or lesser degree, all language areas and skills are being addressed in CALL. The problems only really arise when the computer is required to evaluate user input in some way, especially in less structured, interactive domains. Thus, although the computer can present digitized sound effectively for listening purposes, it cannot evaluate sound input reliably, especially in the form of the spoken voice of the nonnative speaker. Overall, the capabilities of the technology contrive to allow certain kinds of CALL work to be contemplated.

The study selected for in-depth analysis for this section on the language areas and skills was conducted by Dorothy Chun (2001). This study had a number of notable features. It is the only one of the six studies described in this chapter that included correlational research. In this regard, Chun related online look-up behavior with learners' understanding of the text, first in relation to the use of an internal glossary, and second in the use of the external dictionary. Also, the research design in this study catered for individual differences, specifically to record variation in look-up behavior and performance: Chun might easily have omitted this dimension, but she chose not to do so. The individual student's ability level was the main individual difference investigated with regard to learning with hypermedia. For example, one hypothesis investigated whether high-ability students would look up fewer words than would low-ability students. They did, in both the internal glossary and the external dictionary. Finally, the think-aloud protocols and the follow-up interviews also provided a qualitative dimension to the study that assisted in two ways: through the former, by widening the scope of the study to include a discussion of metacognitive reading strategies; through the latter, in cross-checking the findings on the use of the glossary and the dictionary, as well as gauging reactions to the audio narration.

What emerged strongly in this study and others like it was the practical and usable information that readers can take away, either for program design or as a platform for further research. This kind of research can inform many design decisions, especially in relation to the kinds of resources that should be provided, both across the board for all language learners and more specifically with a view to better accommodate the needs and characteristics of the individual learner. In this regard, Chun concluded, "[O]ne of the emerging principles in learning with hypermedia is the importance of individual differences in learning styles and learning strategies. This principle is reflected in the data here in that there was both varied use of the audio narration as well as a mixed reaction to the audio component" (2001, p. 392). The sensitivity to individual difference was one of the strengths of this study on L2 reading and the Web.

Tracking and Labeling Patterns of Learner Behavior: Responding to Difference

Pujolà's (2001, 2002) tactic of allocating labels to groups of learners or users who appear to behave in the same way in particular settings has been used by others, both in technology design more generally and in CALL (see Cooper, 1999; Stockwell & Levy, 2001). In the CALL activities set by Heift (2002), for instance, four "interaction types" were identified among the students according to the way in which they processed the feedback in an ICALL program. Heift labeled the student groups *browsers, frequent peekers, sporadic peekers*, and *adamants*. On the whole, browsers tended to move through the exercises without attempting to answer, frequent peekers requested the correct answer(s) more often than correcting their errors, sporadic peekers used the system help options less than they corrected themselves; and adamants persistently corrected and recorrected their errors without, in the main, asking for the correct answer.

Both Heift (2002) and Pujola (2002) attempted to link the students' language proficiency level with the grouped behavior patterns, although both authors said that more research was needed in this area to firm up any conclusions that they might have drawn. Nonetheless, Heift (2002) suggested in her study that mid- to high performers tended to be adamants, whereas sporadic peekers were mainly students with intermediate language-level skills. She speculated also that beginners took more advantage of system help options.

If student behaviors in these different CALL contexts can be grouped in meaningful and reasonably consistent ways, be it in relation to online tasks or with a tutorial program, the results potentially could be very useful. Such results could lead to improvements in task design and the provision of options and resources that are more responsive to individual differences by offering specific resources for students at the appropriate time and at the appropriate level. In this regard, Cooper (1999), an expert in the user interface design, asserted that we need to aim to develop user "personas" (see also Colpaert, 2004). He continued:

> Personas are not real people but they represent them throughout the design process. They are *hypothetical archetypes* of actual users. Although they are imaginary, they are defined with significant rigour and precision. Actually, we don't so much "make up" our personas as *discover* them as a byproduct of the investigation process. (p. 124)

Cooper provided an extended discussion of strategies for creating personas, and it is worth reading in detail. He argued that designers typically end up with anywhere from three to seven useful personas (Cooper, 1999); however, there should not be too many.

We believe that the identification of patterns of behavior, as shown in the studies by Pujolà (2001, 2002) and Heift (2001, 2002), offers a valuable contributory step in the creation of personas. Personas are composite models. They are informed by patterns of behavior that may embody learner factors such as language proficiency level, first language, educational background, computer literacy, motivation, and so forth (see Dörnyei & Skehan, 2003).

Language-learning tasks may then be designed to suit these personas. In many ways, this approach to design takes a middle path: It neither aims to cater for every individual difference, nor does it provide the same tools and resources for everyone. Instead, it is designed around the personas that themselves are created around a combination of factors that result from individual behavior and characteristics, and salient features of the language-learning environment. It is then not so much a requirement that a program is designed to suit every individual difference, but rather that the individual differences that matter for learning are grouped together appropriately, and then that the help and feedback is designed to respond suitably to these groups. This is also a more realistic goal than trying to customize the program in significant ways to suit every individual, while at the same time it is a more sophisticated response than the one that provides the same help and feedback for everyone with no discrimination at all.

Seeking the Optimal Design

Choosing an optimal design often requires a choice from a limited range of logical alternatives. The study by Jones (2003) is a good example of this kind of work in which the researcher is trying to decide on the most effective form of annotation. This research is comparative, but in a narrow sense where, except for the variable being manipulated (i.e., the annotation type), the research is conducted across a shared context. This is quite unlike the broad comparisons (critiqued earlier in this chapter) in which a researcher attempts to compare composite phenomena such as a Web-based classroom and a "traditional" classroom. Also, even though the Jones study centrally involved evaluation and decisions among alternative design elements, it reached far beyond that. It had deeper implications that have considerable value for the wider audience, and therefore this work moved unambiguously beyond evaluation, as we have defined in this book, to research. This work offered practical outcomes that can be of immediate use to other CALL designers who wish to include multimedia annotations. They can be secure in the knowledge that a certain kind of annotation, in this case visual plus verbal, will be the most effective, and the particular boundary conditions where this finding will apply have been clearly explained. Overall, this kind of study provides a good model for a focused comparison in which the researcher wishes to weigh the value and effectiveness of two or three possible design alternatives.

The Jones (2003) study had a very important qualitative component, too. The students' comments arising out of the interviews shed further light on the use of annotations in CALL design. For example, as well as the students obtaining the lowest scores on all the dependent measures in the control group, the students also had "the lowest opinion" (p. 59) of the listening activity without annotations: In fact, Jones reported that one student referred to this activity as "cruel and unusual" (p. 48). This is an interesting finding. One cannot ignore students' attitudes toward the tasks they are asked to complete, because there is every likelihood that their attitudes will affect their performance. If possible, knowing a student's views toward the task should be considered part of the research. In the Jones study, the transcripts from the interviews showed that a number of students in the control group were very frustrated by the task. They found it too difficult to complete without the assistance of the annotations. It is worth pointing out that when students are regularly working in multimedia, multimodal environments, they come to expect that a range of helpful resources will be made available. Increasingly, aural and visual support is expected. Jones showed conclusively that students performed best when they had access to both visual and verbal annotations.

Here we have a quantitative, statistical finding supported by a qualitative interview response. The qualitative data is most helpful and contributes usefully to the value of the study as a whole. The combination of the two sets of findings serves to strengthen the results and the confidence with which the researcher may draw conclusions from the data. However, qualitative data does not always support quantitative findings.

Student comments from the interviews highlighted some contradictions with the experimental findings. For example, the students' performance was low when only verbal annotations were made available (Group 3); certainly, some students in the verbal group believed that the translations of keywords helped their understanding of the aural passage. Hence, there was evidence of differences within the group that would have been overlooked if only the quantitative statistical data had been collected. In this study, as in others, qualitative data from student interviews provided insights into individual differences. Whatever the broad conclusions of the research study on statistical grounds, adding a qualitative dimension enables the researcher or the designer to be made aware of the variation that often arises as a result of individual characteristics and behaviors. Weinholtz, Kacer, and Rocklin (1995) captured the value of combining methods when they noted "[H]ow ambiguous and misleading results from quantitative studies can be if not supplemented by qualitative data .... Use of supplemental qualitative methods by quantitative researchers can serve as a prudent hedge against obtaining inconsequential or erroneous results" (p. 388).

Finally, Jones' discussion of Mayer's generative theory of multimedia and learning is noteworthy. Sometimes there is a perception that the way that multimedia is designed and orchestrated through its structure and its elements makes no difference as far as learning is concerned. Jones (2003) showed that this is clearly not the case when she maintained:

These results support and extend Mayer's (1997, 2001) generative theory of multimedia learning since the acquisition of new knowledge and comprehension of the aural material was greatest when students processed the text, selected from the relevant verbal and visual information available, organised the verbal and visual mental representations of the annotations into a coherent mental representation, and then integrated this representation into their existing mental model to help them most successfully construct meaning from the aural passage. (p. 50)

Following Jones, we believe we need more research in this area, "to better understand how we can utilise the attributes of multimedia to enhance various aspects of language learning ..." (2003, p. 42).

CONCLUSION

From our examination of the six research strands and example studies highlighted in this chapter, we can make some further observations about research, in addition to those points made at the beginning of the discussion section of this chapter. In planning studies in the future, CALL researchers need to be aware of:

1. Student attitudes and perceptions toward the learning environment, technology, and tasks in different settings. Attitudes may change over time, and longitudinal studies are advantageous.
2. The characteristics of the CALL context, including media, task, and participant characteristics and any significant, known differences between the technology-mediated and face-to-face learning environments.
3. Different task structures that may be more or less effective in different CALL settings (e.g., e-mail, chat with NS, chat with NNS).
4. Different theoretical models that may be used to analyze the same interaction data (e.g., the IA, sociocultural). Using a combination of theoretical models, even when not compatible, may be advantageous.
5. The unit(s) of analysis (e.g., a single interaction, task, activity, learner), which should be carefully defined.
6. Learner training and goal setting, which can significantly affect results; the role the teacher plays can exert a similar effect.
7. Tracking options—new software tools can generate valuable, previously unavailable data.
8. Research designs that evolve around language areas or skills. Such designs can be advantageous for CALL because of the complementary roles that technology can play in language learning.
9. Patterns of individual learner behavior (suitably grouped) in various CALL environments.

10. The need to identify and build different user "personas" in CALL.

In the introduction to chapter 5, we quoted Mitchell and Myles (2004) on the relationship between theory and research in second-language learning. To recap, although Mitchell and Myles accepted the need for cumulative research programs within a theoretical framework, they tended toward a pluralist view of second-language-learning theorizing (see also Jordan, 2004). This is a realistic and workable position, and the one we adopt here in relation to theory and CALL research.

It would undoubtedly be easier to advocate a single research program. This is not advised here, mainly in recognition of the complexity of language and language learning, and the range of problems being tackled in CALL research and design. We are not working in a field where a single theory or theoretical framework is preeminent. To insist on one particular theoretical perspective in the present climate is unrealistic. Instead, given the complexities of language learning, the absence of a grand theory, and the modular approach generally taken toward theory construction and research, we would suggest theory development and research agendas that follow a number of paths simultaneously.

Equally, putting no curb on the number of parallel research agendas would also be problematic, because there would be no pressure to weigh the effectiveness and value of different research agendas. In an educational environment in which resources are strictly limited, this would be irresponsible because there would be no need to make comparisons to locate the best ways forward. Thus, we advise neither a single path nor one that admits an unlimited number of alternatives. Instead, we argue for a mid-way strategy: one that admits a limited number of parallel agendas, while at the same time emphasizing critical comparison and the importance of cumulative research programs. This is, we believe, the optimal practical solution to a complex problem. It is perfectly feasible to proceed along a number of paths in parallel, addressing a limited set of research questions. These paths will be determined from a careful account and critical assessment of existing work in CALL, and thoughtful reflection on the most productive lines of research to take in the future. We have made some suggestions in this chapter that we believe deserve consideration. At intervals, these pathways or research strands should be critically reassessed to ensure that they remain productive (see also Fischer, 1999).

Finally, we must not forget why research is conducted in the first place. It is a mistake to believe that research occurs in a vacuum—in a social environment that is perfectly neutral. Researchers undertake projects for a reason. They may conduct research in response to a successful grant application, in which case there will be certain rules that have to be followed and stakeholders that have to be satisfied. Research may be conducted with a view to gaining a qualification or advancing a career; in such cases, the particular predilections of the supervisor(s) or more senior colleagues will have to be seriously considered. It is a brave research student who enters into a long-term research project that departs from the supervisor's theoretical preferences, which, in turn, are usually determined by his or her own research training and background. We are all shaped by our knowledge and experience, and its limitations (including the writers of this book). The challenge for all of us, students and supervisors alike, is to rigorously and critically assess the options, build on what has gone before, and then choose our direction as thoughtfully and as carefully as we can. In planning CALL research, we can do no more than that.

Chapter 7

Practice

A glance at the literature reveals that the word *practice* is common in the titles of recent books and articles on CALL, and it is often used as an encompassing term for papers that describe the actual use of CALL in the classroom. Egbert and Hanson-Smith (1999), for example, included *Research, Practice, and Critical Issues* as the subtitle of their book: Warschauer and Kern (2000) employed *Concepts and Practice*, whereas Felix (2003a) used *Towards Best Practice*. Clearly, practice is an important dimension in CALL. However, bringing together the issues associated with the practice of CALL in a single chapter—as we have attempted to do here—is a difficult task to achieve, because practice by its very nature is dependent on the individual language-learning environment, the tools that are available to the CALL practitioners, and the expertise of teachers and the learners.

Although there are many possible ways to approach the practice of CALL, we have elected to structure this chapter in terms of the language skills and areas. We have three main reasons for adopting this approach. First, there is a strong history of CALL applications being approached in this manner (e.g., Jones & Fortescue, 1987). Further examples from books and journals include chat and syntactical development (Yuan, 2003), multimedia and vocabulary (Yeh & Wang, 2003), and computer use and writing (Chikamatsu, 2003), suggesting that there is a common tendency to associate technology with the development of specific language skills and areas (see chap. 2, Table 2.1, in this volume for more examples).

Second, we need to bear in mind that technology is not equally strong in all language skills and areas, and there are some aspects that are more amenable to implementation via technology than others. For example, the computer has always been considered strong in the teaching of grammar, vocabulary, reading, and writing, but less so in listening and speaking

(Healey, 1999; Ng & Olivier, 1987). More recent technologies, however, are better able to accommodate these language skills as well. For example, a wider range of practice activities for listening emerged with the capacity to put digitized archives of listening materials on the Web. The technical challenges associated with speech recognition meant that comparatively little appeared in the literature until the hardware and software reached a high level of reliability, and then speech recognition diffused into the population at large and CALL practitioners began making use of it. As a result, we need to constantly reevaluate the capabilities of different technologies; often, this reevaluation is achieved most effectively by considering the appropriate language skill or area in relation to the particular technology in question.

Third, although mainstream language teaching generally advocates an integrated approach to the language skills and areas, this is not necessarily the best approach all the time; sometimes, a more limited focus is beneficial, especially if the student has a particular problem area or wishes to concentrate on improving a particular skill (e.g., academic essay writing). In some cases, an aspect of the language is particularly challenging (e.g., learning script in Japanese or Chinese), and this requires a special focus. Specific technologies used with the relevant language-learning materials can be selected or developed to meet these needs. Some of the time, this compartmentalized approach for CALL complements the second point that certain technologies and CALL materials lend themselves better than others to certain language skills and areas.

Thus, the description part of this chapter provides an overview of the ways in which CALL can be used with the teaching and training in the language skills of listening, speaking, reading, and writing, and in language areas including vocabulary, grammar, and pronunciation. As in earlier chapters, it does this by providing concrete examples from the literature, outlining ways in which these various skills can be taught through CALL, and detailing some of the considerations that must be kept in mind by CALL practitioners.

The practice of language teaching in recent years has certainly placed more of a focus on communicative and content-based learning utilizing a range of authentic language tasks, as opposed to the more form-based language learning of the past. In saying this, however, and as Healey (1999) pointed out, "[C]ommunication does not occur in a vocabulary and grammar vacuum" (p. 116); it relies on the development of other skills. In other words, even though the focus may be on developing communicative ability in the target language, there is still a need to ensure that discrete skills receive sufficient attention.

The wide range of activities available across the language skills can make it quite daunting for new CALL practitioners trying to find an entry point in the design or development of a new activity, and to provide tasks that are suited to their students. The following sections offer an overview of the various ways in which CALL can and has been used in teaching these language skills. Although only a small sample of the ways in which these skills have been developed through the computer is listed here, it supplies evidence for the complexity involved for language teachers to find and/or design tasks that suit their pedagogical needs.

DESCRIPTION

Listening

An essential factor in using technology to teach any language skill or area is that the technology should provide something that is not available through more traditional means. That is to say, materials that are used in the teaching of listening skills should be more than simply audio segments available on the computer in the same way as an audio cassette or CD; they should capitalize on the capabilities of the computer in a way to enhance the learning experience.

A good example of this was provided by Jones (2003; this study is discussed in more depth in chap. 6), who described an environment in which learners of French were given a passage that included only selected keywords shown on the screen, with the remainder of the missing words indicated using ellipses to denote the "flow" of a target aural passage. Through the use of buttons on the display, learners were able to listen to syntactically divided chunks of the passage, or alternatively they were able to listen to any one of the keywords to hear it pronounced. In addition to these audio clues, learners were also provided with visual or text annotations, which allowed them to view pictures or photographs that represented the missing words, or to see text annotations and view their translation. The results of the study showed that students performed better with greater choice of annotation type, and student reactions to the tasks were directly proportional to the number of annotations that were provided. Learners who had more annotation options available to them had a higher opinion of the tasks than did those learners with fewer options, and learners with no annotations rated the activities the lowest. The results yielded from this study support the case put forward by Hoven (1999a), that success in listening with the computer is dependent on the design of the task in terms of catering to individual learner differences, including learning styles and preferred learning modes.

The development of audio- and videoconferencing technologies has also allowed for more communicative types of listening tasks. Studies by Hampel and Hauck (2004) and Wang (2004a, 2004b) have shown the potential of the computer to bring together learners and teachers, even when they are separated geographically. Through these technologies, distance learning courses—which typically have depended on passive listening activities—allow for authentic interaction among participants. Although there are still limitations with the technologies (see chap. 8), they have shown great potential for authentic listening in language classrooms.

Speaking

Speaking is a skill that has usually been regarded as a difficult one to teach through the computer. Although there are aspects of speaking that have been the subject of a number of studies in recent years, such as pronunciation (Kawai & Hirose, 2000; Neumeyer, Franco, Digalakis, & Weintraub, 2000), intonation (Levis & Pickering, 2004), and vowel contrasts (Carey, 2004; Wang & Munro, 2004), studies that focus on speaking itself as a skill have not featured heavily in the literature (e.g., Holland et al., 1999). As far as CALL is concerned, it is advantageous to consider speaking and the computer in three ways: tasks that require the learners to speak, tasks that peripherally assist the skills required for speaking through focusing on other skills, and tasks that require the computer to recognize and respond to language input.

The first of these is conducted through computer-based interaction tools, such as video- or audioconferencing, in which learners engage in speaking with teachers or native speakers (as discussed in chap. 4) and is most common in distance education settings (Hampel & Hauck, 2004; Rosell-Aguilar, 2004). The benefits for oral proficiency are obvious, because the learner is directly involved in the act of speaking with teachers, native speakers, or other learners. Although there are obvious differences in communication as a result of the intervention of the technology, the nature of conferencing is sufficiently close to consider it similar to speaking in face-to-face contexts. Despite this, audioconferencing is faced with its own difficulties. Stevens and Hewer (1998), cited in Kötter et al. (1999, p. 58), argued that the "synchronous nature of the medium, along with the lack of eye contact and visual cues resulted in the need for tutors to manage the audio conference very carefully and sometimes caused students to over-prepare to the extent that they read out prepared answers." Stevens and Hewer also argued that the public nature of conferencing—and this is applicable in both video- and audioconferencing—causes teachers to be faced with difficult decisions regarding the best way to deal with error correction. This, combined with confusion expressed by learners regarding turn taking and the ability to "hide" in anonymous interactions that often happens in multi-participant audioconferencing environments (Kötter et al., 1999), adds a complexity for speaking in conferencing that is not present in face-to-face situations.

The second way in which speaking skills may be developed through the computer contrasts with the first, in that interactions occur through text-based means, and do not require the learners to physically "speak" with each other. There is some preliminary evidence suggesting that other nonverbal forms of CMC also have the potential to improve L2 speaking skills. A good example of this was provided by Payne and Whitney (2002), who described a study in which learners involved in synchronous CMC classes were compared in their L2 oral development with learners who received only face-to-face instruction. The subjects involved in the study were students of Spanish who were required to communicate with each other and the instructor through synchronous chat for one half of their classroom instruction, meaning that the CMC mode comprised a significant proportion of the total time spent on the students' Spanish instruction. The results of the study showed that those students who were involved in the synchronous CMC interactions demonstrated significant improvement in their oral proficiency as measured by a modified version of the ACTFL OPI (Oral Proficiency Interview). Less convincing results were seen by Abrams (2003b), who suggested that students of German involved in synchronous or asynchronous CMC failed to show any significant improvement lexically or syntactically when compared with students who received only face-to-face instruction.

The third way that the computer has been used for development of speaking skills requires the computer to recognize and respond to oral output from the learner. Two examples of this are described here. In their *Virtual Conversations* program, Harless et al. (1999) investigated military linguists completing a course in Arabic to maintain their language level. The program required the students to interact with Iraqi characters in an interrogation scenario. The results showed significant improvements in speaking (and, incidentally, listening) scores across the intensive week-long study, and reactions to the software were very positive. A similar concept is presented in a program designed for beginner Japanese students entitled *Subarashii*, by Bernstein, Najmi, and Ehsani (1999), in which learners are faced with various situations that they must deal with in Japanese, including greeting someone for the first time and buying milk from the convenience store. Although no measures of oral proficiency were taken in this study, reactions to the program were very positive. Both programs relied on pattern matching, in which students were required to produce utterances from either a presented choice of options, or options that had been predicted by the designers. In this sense, the type of task bore some similarity to pronunciation training in that the interactions were not open as such, but instead relied on the learners producing fixed responses to the given situations. In cases in which incorrect responses were given, or the pronunciation of parts of the utterances was not recognized by the computer, appropriate responses were made by the computer. Having the computer interact with the learner in completely open interactions is still proving difficult to realize, as is described in the following chapter.

Reading

Reading through the computer has been thought to have various benefits associated with it; namely, wide access to authentic materials instantly through the Internet, the ability to add hypertext for accessing vocabulary or grammar explanations, greater access to reading authentic communication materials, inclusion of multimedia in conjunction with plain text, and the ability to control for reading speed. The Internet has proven to be a rich source of reading materials that, if used well, can motivate students as well as contribute to linguistic development and cultural understanding (Taylor & Gitsaki, 2004). There are, however, a number of concerns that have been raised about the use of the Internet, including the difficulty in locating appropriate materials due not only to the amount of information available, but also to the varying levels of appropriateness, linguistically and socially. Provided that teachers are aware of the possible limitations and dangers, it is possible to find a great deal of material that is adaptable to classroom situations (Dudeney, 2000).

Hypertext links allow teachers to place extra information such as translations, examples, or even multimedia links to pages (Gettys, Imhof, & Kautz, 2001). Through hypertext, learners are able to click on links while reading to find information about certain words or phrases without the need to stop the reading process completely to look up unknown items in a dictionary. Another way in which the computer has contributed to the reading process is through CMC, particularly asynchronous forms such as e-mail or BBS. Although the use of authentic interactions provides meaningful reading for learners, one shortcoming noticed by researchers (e.g., Moran, 1991) is that e-mail interactions are prone to becoming an exchange of monologues between the participants, in which learners compose messages without reading the messages received from their partners. Learners can be instructed to respond to the content of e-mails, but in cases in which there are multiple topics in a single e-mail, it is still not uncommon for learners to fail to respond to each of the topics (Stockwell, 2003a).

Writing

Writing through CALL has taken a variety of forms. One of the earlier but enduring forms is via a word processor, by which learners have access to a

range of "advanced writing tools and a facilitative environment for generating ideas and producing text, both as drafts and finished copy" (Pennington, 2004, p. 71). The reasons Pennington cited for the word processor's
suitability to writing is its capacity to ease the "mechanical" processes of
generating text, which is largely attributable to its ability to allow revisions
through deletion, addition, substitution, and moving large blocks of text.
For the most part, word processing still appears to be one of the strongest
supports for composition writing.

Another way in which writing has been encouraged through the computer includes authentic communication through some form of CMC. Generally speaking, asynchronous CMC such as e-mail is thought to be better
for development of writing in a second language, due to the fact that the
language produced bears greater resemblance to written language than
that produced via synchronous CMC, such as chat (see chap. 4 for further
discussion of this issue). There is growing evidence that asynchronous CMC
provides learners with input models for development of writing in a second
language. Davis and Thiede (2000) showed that learners of English involved in asynchronous discussions demonstrated style shifting, reflecting
the language produced by their interlocutors, both in terms of stylistic emu-
lation and in syntactic complexity. Regarding the use of e-mail, Fotos
(2004) found that learners exchanging e-mails with their tutors experienced proficiency gains measured by the reading and vocabulary components of a TOEFL test, and Li (2000) noted that students involved in
authentic e-mail interactions tended to produce texts that were more
syntactically and lexically complex.

Other ways of using CALL to assist with second-language writing have
also featured in recent literature. Chambers and O'Sullivan (2004), for example, discussed an environment in which advanced learners of French
consulted corpus data on topics similar to those that they were writing about
in order to improve their writing. The authors discovered that learners
made significant numbers of changes to their writing on the basis of the corpus information. Corpus data are commonly used in the teaching of grammar and vocabulary as well, and are described in more detail later in this
chapter.

As part of the writing process, then, CALL can be used to provide ways to
expose learners to individualized native speaker or teacher input, to think
about alternative ways of expressing ideas, and to easily revise and edit the
writing.

Grammar

As with other language skills and areas, there is variation in the approaches
used by teachers to teach grammar. Grammar and vocabulary often appear

together, and although there are several studies that focus on one or the other of these skills individually, complete separation of the two skills is often difficult (Kempen, 1999; Labrie, 2000; Loucky, 2002; Tschichold, 1999). Grammar teaching in CALL generally takes one of three forms: grammar tutorial exercises, learner-centered grammar instruction, or communicative grammar instruction.

Grammar tutorial exercises through CALL are more typically associated with the "traditional" CALL exercise, and were perceived as one of the greatest uses of CALL in the early days. Although a highly valid and useful application of CALL, drill-based grammar activities, which comprise a significant proportion of grammar tutorial exercises, appear to have been the target of criticism in recent years, and many teachers have opted for the more communicative type of activity described later in this section. Despite this general movement away from drill-based grammar instruction, it is perhaps ironic that many people new to the field consider this type of activity to be the essence of what CALL is. Drill-based activities most certainly still have their place in the language curriculum. As Healey (1999) explained, there are learners who prefer to learn rules before attempting to use new language forms in more communicative settings, and many learners are able to gain a "sense of security" in being able to do activities in which the required answers are clearly defined (as opposed to completing a communication act, for which there is not necessarily a single correct response). Despite the inclusion of other, more communicative activities through CALL in recent years, grammar-based tasks created by teachers for their individual language-learning environments using authoring software such as *Hot Potatoes* or other similar tools are a component of many language-learning environments (Chan & Kim, 2004). Drill-type grammar activities are generally in the form of more traditional types of gap-filling tasks or multiple-choice questions, but they are certainly not limited to these. In addition, there are a number of grammar tutorial activities that are not drills, in that they are not mechanical and involve conscious reflection on not only form but also on meaning and usage (see Hubbard & Bradin Siskin, 2004, for a discussion of practice and drill activities). Grammar instruction may also be assisted through the use of parsing software, which checks for the grammaticality of sentences entered into the program and provides feedback on the basis of the learners' responses (Dodigovic, 2002). This type of activity places more emphasis on the learners to produce language output more actively, but can be very difficult to produce on the part of the CALL practitioner, making it more limited in its application.

Learner-centered grammar instruction places greater emphasis on the learners to deduce the rules of the target language for themselves. One means of doing so that features regularly in the literature is the use of concordancing and corpora. A concordancer is a piece of software that arrays the occurrences of a given lexical item or expression from a corpus of language data in context so as to allow for comparison of the usage of the item or expression (often referred to as "key word in context," or KWIC). Also used extensively in vocabulary-learning tasks, the main tenet of using this type of approach—called a "data-driven approach" to language learning—is that the responsibility is placed on the learner to make their own comparisons of the way in which target expressions are used, and thus to formulate rules on their own. Concordancing can be monolingual, in which the learner examines only the target language data (e.g., Curado Fuentes, 2003), or it can be bilingual or *parallel* (St. John, 2001; Wang, 2001), in which the learner can see examples of first- and second-language data together and draw conclusions based on this comparative information. Depending on how they are used, concordancers are also frequently employed in the development of second-language vocabulary, in that they demonstrate target vocabulary or expressions in a range of naturalistic contexts, providing the learner with extra information that may lead to a greater understanding of selected items and how to use them (see Chambers, 2005, for a discussion of concordancing in second-language learning).

Communicative grammar instruction also has an important role to play in second-language teaching, because it allows learners to go beyond the scope of language included in grammar tutorial exercises and learner-centered grammar instruction activities designed by the teacher. Although both of these have been used successfully in teaching, they are limited in that they are only as comprehensive as the range of activities and explanations provided by the teacher, or the size of the corpus linked to a concordancer. The only way in which learners are able to have access to "unlimited" language input is through authentic communication, be this with native speakers or with other learners of the language.

The choice of the appropriate form of computer-mediated communication is a difficult one, and different types of CMC allow for emphasis on different aspects of language learning. As described earlier (and in more detail in chap. 4), asynchronous forms of CMC are potentially better suited to grammatical development, although, as Sotillo (2000) noted, mistakes are present in asynchronous forms of CMC as well. The broken nature and abbreviations that are often characteristic of the language produced in synchronous chat mean that it is less suited as a medium for promoting accuracy in the use of the forms of the language. Clear and focused e-mail tasks have the potential to lend themselves well to grammatical development, in that learners have both time to edit messages before sending them, and time to read through messages using dictionaries or other resources during message composition. Stockwell and Harrington (2003), for example, suggested that learners were able to demonstrate significant improvements in syntactic complexity and accuracy over a relatively short period of time while involved in e-mail interactions with native speakers of Japanese. Analysis of the interactions revealed that learners used the input from their native speaking interlocutors in one of two ways. First, there were examples of self-correction in which learners were exposed to the correct form provided by their native-speaking interlocutors following an error, either through explicit correction on the part of the native speaker, or through learners' noticing their own errors and correcting their language output accordingly. In addition to this, there were several cases in which learners recycled constructions used by their partners in subsequent messages.

Vocabulary

Studies into vocabulary acquisition through CALL typically fit into one of two main approaches. The first of these sees vocabulary as a skill to be taught explicitly, and thus presents vocabulary to learners in a style that lends itself directly to the acquisition of the target vocabulary items. Ma (2004) identified three types of CALL programs that deal with vocabulary learning that would fit into this first approach: complete multimedia packages that include a vocabulary learning component, programs that are made up of written texts with electronic glosses, and programs dedicated to vocabulary learning.

The first of the three types described by Ma (2004) is perhaps the most commonly found type of vocabulary software, and the software review sections of any of the major CALL journals frequently contain examples covering a large range of different languages. The majority of these software packages include components for learning vocabulary, and are varied in the approaches taken. The second type includes written texts with electronic glosses that can be accessed by learners in the process of reading. One of the most common ways in which glosses are provided in CALL-based reading activities is through provision of hyperlinks. In a study by Laufer and Hill (2000), target words were linked to definitions in the second language, translations in the second language, and audio recordings of the words. Although there was significant individual difference, they found that use of the electronic glosses contributed to higher recollection of the target words. They argued, and were supported by De Ridder (2002), that the hypertext marking of the items contributed to making the items more salient to the learners, and thus may have played a role in assisting in the acquisition of the items as shown in posttests. A further addition that CALL enables is to provide not only definitions, translations, and audio annotations, but also pictures and video segments, which have also been linked to enhanced acquisition of vocabulary (Nikolova, 2002; Yeh & Wang, 2003). An example of the third type—programs dedicated to vocabulary learning—was provided by Groot (2000). He described a study in which groups of vocabulary items were systematically presented to learners in four stages: deduction, in which the learners were required to deduce the meaning from context; usage, in which learners consolidated knowledge of word meaning through usage exercises; examples, in which words were presented in authentic contexts; and retrieval, in which learners had to type in the correct word for a given sentence.

The second approach sees vocabulary as something that is to be acquired peripherally while the student is engaged in an authentic task, such as looking at pages on the Internet or through communicating through some form of CMC. This differs from the reading texts with electronic glosses in that in the earlier case, through the provision of an electronic gloss, it is still the intention of the teacher to have the learners acquire vocabulary. Through authentic communication or reading tasks, although the teacher may desire for his or her learners to naturally acquire—or even consciously study—vocabulary through the process of the writing, the primary objective is still to communicate meaningfully with an interlocutor. There is evidence that communicating with native speakers through CMC has the potential to lead to lexical development. Toyoda and Harrison (2002), for example, gave examples in which there was explicit correction or explanation of lexical items on the part of the native speaker.

Developments in Web-based technologies have made the use of authentic materials more accessible to students as well. Hosoya (2004) wrote of a system entitled *WebOCM*, which has the capacity to change every word in a Web-page accessed by students through the system into hyperlinks to German-Japanese glosses and audio representations of the words. Through systems such as this, teachers have the capacity to allow students to search for information in pages in any field that they like, rather than trying to second-guess the type of areas in which students may be interested.

Pronunciation

With the broad acceptance of the communicative language-teaching approach, pronunciation underwent a period of reduced emphasis in the language-teaching curriculum, and although today there is a general increased awareness of the importance of teaching pronunciation for successful language processing and learning (Cook, 2001), there are still questions as to how it can best be integrated (Burgess & Spencer, 2000). Despite the fact that many language teachers are aware of the importance of teaching pronunciation, the practice of pronunciation instruction is often given minimal time within the language class due to the logistical constraints involved in having the teacher repeat the required focus points as many times as is needed for each individual student. Using the computer to teach pronunciation has the potential to alleviate the burden placed on the teacher, in that students who learn at different rates are given the freedom to have required sounds repeated to them as many times as is necessary (Healey, 1999); and there is a growing body of research into computer-assisted pronunciation training (e.g., Hardison, 2004; Kawai & Hirose, 2000; Wang, Higgins, & Shima, 2005).

There are two main approaches to teaching pronunciation in CALL. The first of these is one-way, in which instruction on how to produce specific sounds is provided either in the shape of graphic annotations or through the use of illustrations, photographs, or multimedia. This type of instruction was more popular in early CALL, but has essentially now been completely replaced with the second form, in which computer pronunciation tutors "listen" to learner output and provide feedback based on this output.

An excellent example of a package that provides more detailed learner feedback was given by Tsubota, Dantsuji, and Kawahara (2004) for Japanese learners of English. The software identifies the aspects of English pronunciation in which the learners are experiencing difficulties, specifically searching for 10 areas predicted as being problematic for Japanese learners. After identifying the areas in which the students require more practice, the software then automatically provides feedback and practice in those areas where errors were detected. Software for pronunciation instruction need not be expensive or limited to high-tech laboratories. Brett (2004) described a project in which real-time feedback is provided on vowel pronunciation of students of English using an open-source (free) software package entitled *PRAAT*. The *PRAAT* program allows for speech analysis and synthesis, and is designed to be quite customizable to suit a range of individual needs and requirements. In addition, it runs on a variety of platforms, making it accessible even to CALL practitioners with less than ideal financial and hardware resources.

Despite the greater technical sophistication of feedback-providing pronunciation software packages, the question of which type of feedback is more effective in improving student pronunciation is still contentious, as illustrated in a study by Hew and Ohki (2004). They compared animated graphic annotations and immediate visual feedback on the development of Malaysian learners' pronunciation of Japanese. The graphic annotation system used symbols to represent long sounds, double consonant sounds, aspiration, and voicing of the sounds to be produced, thus increasing awareness of features of the words the students were practicing. The visual feedback system showed graphic representations of the sounds recorded by the learners alongside representations of native-speaking models to allow learners to compare and identify problem areas in their pronunciation. Although the results do not conclusively confirm the superiority of one method over the other, both contributed to improvements in pronunciation development by the learners. Hew and Ohki (2004) suggested that learner style could be a contributing factor here, with animated graphic an-notations helping learners to build visual images, and immediate visual feedback providing individualized information on the pitch, duration, and frequency ranges of their output.

This section has briefly described the variety of ways in which the com-puter has been applied to teaching various language skills and areas, in-cluding reading, writing, listening, speaking, grammar, vocabulary, and pronunciation. Although it is far from exhaustive in supplying examples for each of these areas, it provides a cross-section of the options that are open to language teachers who are currently using, or planning to use, technology in their language teaching.

The practice of CALL is complex. Those who use CALL in the classroom need to have a clear idea of what they want to achieve, the technological op-tions available to them, and also the backgrounds and goals of the learners. The discussion section that follows examines what is required for successful application of CALL to a learning environment from the perspective of having clear objectives, knowing the technological options and their peda-gogical implications, and knowing the students abilities, goals, and percep-tions related to different types of CALL. It continues with a discussion of some of the points to bear in mind when developing a CALL task, how to work with constraints, and how to maintain the best balance in implementing CALL in language-learning situations.

DISCUSSION

Knowing What You Want to Achieve

Successful use of CALL depends heavily on teachers having a clear idea of what they want to achieve in the classroom. Questions need to be asked about language-learning objectives and what language skills or areas are to be taught before the choice of technology suitable for realizing these objectives can be made. It is all too easy to embark on using CALL in the classroom before clearly defining what it is that is to be achieved, both in terms of the learning goals and the role of the technology in achieving them. One of the common traps when making decisions about introduc-ing computers into a language-learning situation is to start by developing language tasks based solely on the functionality of the computer, and care must be taken to ensure that teachers don't allow themselves to be led only by the technology available at a given time. In saying this, however, there is a need to understand the empowering and limiting features of any tech-nology, and what the technology can achieve in relation to the language skills and areas in order to make informed choices about how to imple-ment a CALL component.

As discussed in chapter 2, the type of tasks that are used will depend very much on the approach adopted by the course developers. Courses that predominantly revolve around specific content will obviously require different types of tasks than will courses that are more skills based. It follows, then, that teachers and course developers must have a clear idea of the learning objectives of the course and develop the CALL tasks accordingly. There is a great danger in using CALL without first clearly defining these objectives, and we would question the value of tasks about which the learners (and often the teachers) are left wondering why they are doing them. Chapelle (2001) advised caution regarding CMC-based tasks; for example, she argued that although CMC obviously provides opportunities for learners to interact with native speakers or other learners of a language, simply bringing learners together in unfocused communication is unlikely to lead to any significant developments in the target language. CALL-based tasks are only likely to be effective if the purpose of the course has been clearly defined from the outset, and the learners themselves have been made aware of the requirements and the expectations placed on them in achieving the task.

Healey (1999), stated, memorably, that "technology alone does not create language learning any more than dropping a learner into the middle of a large library does" (p. 136). Simply providing the learners with technology in the hope that this will in itself lead to development of language skills on the part of the learners is ambitious, to say the least. The way in which technology is used, and indeed the very choice of what technology to use in any given environment, depends on a clear vision of language-learning goals and technological options.

Knowing the Technological Options and Their Pedagogical Implications

CALL practitioners are faced with decisions about what type of hardware and software should be adopted in achieving specific learning objectives. Being aware of what technological options are available in teaching each of the language skills and areas goes a long way toward ensuring that these decisions are made appropriately. For example, if the main goal of a class is development of listening skills, then teachers need to know what types of technology are best suited for listening, or if a class is to help learners with learning vocabulary, then teachers need to consider what technologies are appropriate. This knowledge allows teachers to employ the language-learning technologies in a way that enables the students to achieve their learning goals.

Different technologies for language learning automatically lend themselves better to some types of tasks than to others. The concept of affordances has been raised at various points in this book, and it has been used to describe the way in which technology may facilitate or constrain language learning. It is natural that such affordances also have an implication on the pedagogical applications of different technologies. Synchronous CMC such as chat, for example, tends to be more lexically focused due to the limited amount of time that it provides for learners to think between receiving input and responding to it (Skehan, 1998). Hence, as described earlier, asynchronous forms of CMC such as e-mail lend themselves better to development of syntactic skills (e.g., Sotillo, 2000). The language used in chat tends to be less well formed than the language produced in other forms of CMC. Thus, in a language-learning environment in which the focus is more on development of grammatical accuracy, chat would be unlikely to be the ideal choice of technology. In this way, it is important that course developers consider the individual features of the technology, and how these features can be used to help achieve the pedagogical goals of the given course.

With technology it is possible to add extra dimensions to the learning environment that are not present in more traditional settings. *Multimodal learning* refers to learning that involves a combination of media types, including face-to-face interactions. Through multimodal learning, it is possible to counter some of the disadvantages associated with any communication mode used separately. Synchronous CMC requires all participants to be online at the same time, whereas in asynchronous forms, messages can be left for recipients to respond to at a more convenient time. Thus, even when communication participants are spread across multiple time zones, through using a variety of media it is possible to envisage an environment in which shorter periods of time are set aside for synchronous communication when it is suitable for all involved, yet communication can still be carried out through other modes to increase the amount of input and output. Examples of this were provided by Möllering (2000) and Hampel and Hauck (2004), who showed, in rather different ways, how the different media or modes can be applied to different language-learning situations and objectives. In Möllering's study, although the majority of communication in the target language between learners occurred through a bulletin board, e-mail was used for assignment submission and for requesting information. In the study by Hampel and Hauck (and also described in Hampel & Baber, 2003), the VLE *Lyceum* allowed for real-time audio communication among participants, and at the same time included a graphical interface for visual exchange of ideas.

Providing learners with a wider range of modes of communication does not guarantee that they will be used effectively. Ligorio (2001), in a study of learners involved in a cross-institutional intercultural collaborative project employing both synchronous and asynchronous as well as text-based and visual forms of communication formats, found that this variety was only appreciated and used by the participants when the learners had reached a particular level of awareness of the technical and cognitive functions of each of the formats. Initially, learners mainly communicated through chat to establish interpersonal relationships, and only after they became familiar with the range of tools available, through both self-exploration and teacher guidance, were they able to make use of several formats in achieving their goals. The fact that learners relied more on synchronous forms of communication in preference to asynchronous forms in the early stages is not surprising. As described in chapter 4, the sense of distance associated with different modes of communication varies greatly (Weininger & Shield, 2004), with much more "closeness" felt by participants in synchronous forms of communication compared with participants in asynchronous forms. It seems natural that synchronous communication is more likely to lead to a quicker establishment of relations than asynchronous communication, in part due to the faster turnaround time between messages exchanged.

Knowing Students' Abilities, Goals, and Perceptions Related to Different Types of CALL

When embarking on the use of CALL in a language-learning context, it is important to bear in mind that the needs, goals, and perceptions of the learners may vary greatly from those of the teacher as well as from each other. Learners enrolling in courses will most often come from a range of backgrounds linguistically (i.e., their experience in both the first and second language), so it follows that these learners will also bring with them different experiences and abilities with regard to technologies. In classes that do not include technology, it is expected that most learners involved in second-language learning will have a basic knowledge of how to use the tools that they will need in the classroom. When technology is introduced, however, these expectations are no longer valid, because it is feasible that some students will come into the classroom with extensive technological skills, and others with virtually none at all.

CALL practitioners thus need to be aware of what skills the learners are bringing with them into the learning environment, and what type of training is necessary above and beyond the pedagogical goals of a course. Hubbard (2004b) presented the problem clearly:

> A fundamental quandary in CALL is that learners are increasingly required to take a significant amount of responsibility for their own learning, whether that learning is taking place through the programmed teaching presence in tutorial software or the unstructured spaces of the World Wide Web. They are expected to do this despite the fact that they know little or nothing of how languages are learned compared to an appropriately trained teacher. And

they are expected to do this within a domain—that of a computer—that is still relatively unfamiliar as a language-learning environment to most of them. (p. 45)

There are two keys issues that Hubbard pointed out in this statement. The first is that many learners who are introduced to CALL are not equipped with the skills required to learn a language, and yet are expected to undertake language-learning tasks, in many cases without the assistance of a teacher. The second is that many of these learners lack the skills needed to utilize the technology through which they are trying to learn. Thus, Hubbard argued for the importance of sufficient *learner training* in both of these areas to ensure that the learners are able to undertake the activities expected of them. In any CALL environment, the responsibility to ensure that learners are aware of what they need to do, why they are doing it, and how they are to do it lies entirely with the teacher. In addition to this, there is an underlying requirement for teachers to understand what it is that learners *need to know* to use CALL successfully as a part of their language-learning studies. Although the pedagogical aspects such as the requirements of a particular task or activity tend to be more generalizable, aspects that are specific to a particular group of learners—and, of course, individual learners—are far more difficult to isolate. Knowledge of the backgrounds of individual learners can go a long way to reducing the burdens placed on them. Stockwell and Levy (2001) found, for example, that students involved in e-mail exchanges who had no prior experience with e-mail or word processing generally produced significantly less output than did learners with more experience. If information such as this can be determined in advance of a project, it is possible to incorporate training in using word processors or e-mail software before learners embark on the main part of the project, enabling them to maximize the benefits of the experience.

Simply having learners take part in even the best-conceived CALL activities without the necessary training in the skills required to perform them will obviously lead to poor image formation and will reduce the potential value of the activities in terms of both what the students may achieve in the short term and in the opinions they form regarding future use. In other words, great care and attention must be paid to how learners are first introduced to computers in a language-learning environment. Assumptions regarding the skills that learners bring with them into the classroom are not always correct (Stockwell & Nozawa, 2004), and presuming that learners have skills that they do not actually possess can cause frustration on the part of the learners and the teachers alike. Although the number of learners without any kind of computing experience at all is decreasing, it is still rather optimistic to believe that younger generations of learners are "by nature or default, pre-conditioned and pre-disposed to-ward computers, multimedia or computer assisted language learning environments" (Hémard, 1999, p. 223). Thus, it is very feasible, and indeed quite likely, that there will be a significant number of learners within any language-learning group who have never used CALL before. Computing skills do not automatically translate into an ability to use CALL materials, and although students may be quite proficient with using computers, this does not guarantee success with CALL materials without the necessary instruction in how to use them.

If the first experience that these students have with CALL materials is that they are difficult to understand and require skills that are beyond the students' technical competencies, the resulting negative perspective toward them is likely to be difficult to overcome, and could adversely affect students' future use of CALL materials. In many cases, problems of negative image formation can be relatively easily avoided through the teacher gathering sufficient information about the backgrounds of the learners, and then attempting an activity through the eyes of the learners, taking into consideration the skills that they do or do not possess. As with any other language-learning task, the learners need to be made aware of the reasons they are doing a particular task and exactly how they should go about it. Depending on the complexity of the task, this training may take only a few minutes, or it may require a number of steps over several classes. In designing any curriculum that includes a CALL component, the amount of time that must be dedicated to instruction in how to use the CALL materials and any extra skills or competencies that are required to use the materials effectively must be considered from the outset.

It is also possible that learners will come into the classroom with their own preferences for learning that may be independent of their actual experiences and skills. They may have varying perceptions of different functions of the computer, as well as preconceptions as to the role of the learner and the role of the teacher in language learning. The actual function of different computer-based applications as a means to second-language learning may conflict with what learners perceive these applications to be. For example, learners who consider that chat is a way of casually communicating with friends or anonymously with other people may find the use of chat as a language-learning tool peculiar, and hence may have difficulties in adjusting to a role that they find uncharacteristic. Examples of this have been seen in several studies (e.g., DiMatteo, 1990), in which learners essentially ignored the assigned task in favor of inappropriate communication in both the target and native languages. Adjustment time for learners may be necessary, as well as explanations of and training in how different technologies can play a role in the students' language learning.

Designing CALL Materials

Design of CALL materials is a far more complex process than simply compiling a list of Web sites or computer-based resources that loosely follow the content of the course in which they are to be used. As we indicated in chapter 2, the point of departure for the design of a language-learning task is critical, and heavily influences the shape of the materials that are produced. It should be kept in mind that CALL materials will very often make up only one component of a language-learning task, and learners may well spend a significant proportion of time away from the computer. Thus, when designing CALL materials, it is essential to remember what the pedagogical objectives in selection or design of a task are, how the task relates to other tasks, and what role the computer plays within the task.

As Chapelle (2003) described, technology-based task design needs to take into consideration several issues, addressing what is to be done, who is to do it, how it is to be done, and what happens to the outcomes. These questions—reproduced in Table 7.1—are highly valid, and can help to guide CALL materials designers to focus the materials to suit the learners who will use them. The topics and actions refer to exactly what it is that the learners are expected to achieve and the processes that the learners must use in order to achieve these goals. These goals should be considered in terms of their integration at many levels (see chap. 9 for a discussion on integration).

TABLE 7.1

Framework for Technology-Based Tasks (Adapted from Chapelle, 2003)

Task Aspect	Task Feature
Topics and actions	What is the task goal?
	What are the topics?
	What processes are used to develop the topics?
	How cognitively complex are the topics and processes?
	Where does the task take place?
Participants	Who are the participants?
	What are their interests with respect to language learning?
	What is their experience using technology?
	How many participants are engaged?
	What is the relationship among the participants?
Mode	What are the modes of language use?
	How quickly must the language be processed?
Evaluation	How important is it to complete the task and do it correctly?
	How will the learners' participation be evaluated?

From *English Language Learning and Technology,* 2003, pp. 138–139. With kind permission by John Benjamins Publishing Company, Amsterdam/Philadelphia. www.benjamins.com

Design of the material needs to take into account the cognitive complexity of the topics (i.e., how personal, how controversial, or how technical) and the processes (i.e., the type of discourse for accomplishing a particular function), and where the task will take place. Teachers need to consider whether the task will occur in the classroom or, alternatively, if it will occur outside of class time in a self-access center or at home. Considering these factors helps to ensure that the necessary facilities are in place for learners, and to determine whether or not the task will be appropriate for the learning environment.

The importance of knowing the participants has been discussed in some detail in the previous section, and the points raised by Chapelle (2003) complement those stated earlier. Apart from knowing the background of the participants in terms of language-learning goals and experience with technology, there is a need to know how many participants are to be involved and what the relationship is among these participants. The number of participants can play a role in the effectiveness of a task, particularly when communication tasks via CMC are concerned (Stockwell & Levy, 2001). The relationship among the participants can affect the dynamics of their interactions as well, that is, if the learners are already quite familiar with one another then there is the possibility that group work is easier to achieve, at least in the early stages. In saying this, however, knowing who the other participants are can also have an inhibiting effect, by which students may be concerned about making mistakes in front of others, or revealing personal information. The potential anonymity afforded by CMC has long been claimed as an advantage of this mode of communication, as shown in studies such as those by Warschauer (1996), in which learners reported that they felt they could express their opinions more freely, comfortably, and creatively through the electronic discussion, with improved thinking ability and reduced stress compared with face-to-face discussion.

As Goodfellow (1999) asserted, there is a need to consider the development of CALL materials from the learners' perspective, thinking of their preferences and how they view specific technologies. A significant proportion of the early research in the CALL literature—and even more recently—investigated student reactions and perceptions to different types of CALL. Hémard (1999), for example, described a case study in which students evaluated the design of a hypermedia CALL package. Jones (2003) compared students' feedback to different types of computer-based listening activities, whereas Pérez (2003) elicited views of synchronous versus asynchronous CMC. Hughes, McAvinia, and King (2004), on the other hand, administered questionnaires to determine student preferences for language-learning sites on the Internet. The current body of research into learner perceptions of CALL materials provides a good foundation for future design, and materials developers would benefit from taking advantage of this information.

The use of CALL assumes that learners possess some degree of autonomy. Although of course it is not always the case, it is still common to see learners plunked down in front of computers and told to work on given tasks for a period of time, and there is often an expectation that learners will continue to focus on the tasks with little or no teacher intervention. Learner autonomy is an attribute that is notoriously difficult to define. It is not constant, but rather is both culture- and task-specific, as well as dependent on each individual learner's motivation. The ways in which learners will approach different types of CALL will depend very much on these factors as well. There are dangers in assuming that all groups of learners will automatically have the skills and discipline required to undertake CALL unsupervised from the outset. For the most part, learners need to be gradually trained and guided to be able to take increasing responsibility for their own learning (see Skehan, 1998, for a discussion of equipping learners for autonomy).

The next aspect that Chapelle (2003) addressed was that of the modes of language use. Here, she referred to the means through which the learners will participate; that is, through face-to-face communication or through some form of CMC, such as e-mail, chat, or even audio-conferencing. Synchronous forms of CMC impose a heavier burden on language learners, and thus are more suitable for higher-proficiency learners than are asynchronous forms, in which learners have more time to process input and produce output. Design of CALL materials needs to take this into consideration, and the time taken for learners to process the input should be factored into the guidelines associated with the materials.

The final issue that Chapelle (2003) raised was that of evaluation. Learners are generally quite conscious of the expectations placed on them, and are concerned as to whether they are completing tasks to the satisfaction of the teacher, particularly when assessment is involved. Learners need to be clearly shown the importance of completing a task and be made aware of whether or not there are any implications for completing it correctly or not. In other words, if the expectation of learners is to complete a task with total accuracy, then they are more likely to spend greater amounts of time on each individual component. If no significance is attached to accuracy, but is rather attached to whether or not the task is completed within the assigned time, it is likely that much of the task will be given less than sufficient attention. Just as teachers have expectations as to how learners should to use materials that are created for them, it is only natural that due attention be given to informing the learners how their participation is to be evaluated.

Working with Constraints

Although CALL has the potential to enrich a language-learning environment, effective use of CALL in the classroom requires practitioners to be aware of the constraints that the choice of certain technologies brings, and how to deal with these constraints. Teachers and administrators associated with technologies for learning are confronted with problems that have the potential to limit or even completely stop the use of these technologies.

Some problems are perceived as global in the introduction of technologies for education. In a study of primary and secondary schools across 24 countries, Pelgrum (2001) related what the school principals and technology experts perceived as the major obstacles to realizing their schools' computer-related goals. Although the study was not targeted specifically at the second-language-learning context, the issues raised are of relevance to CALL facilitators as well. The major obstacles that were listed by the respondents are shown in Table 7.2, sorted by the average percentage of respondents across the countries. Although the top 10 obstacles all attracted responses in excess of 50% of the population tested, according to the results of Pelgrum's study, the largest obstacles to the use of ICT in the classroom

TABLE 7.2
Obstacles to Realizing Schools' Computer-Related Goals
(adapted from Pelgrum, 2001)

Obstacle	Percentage
Insufficient number of computers	70%
Teachers lack knowledge/skills	66%
Difficult to integrate in instruction	58%
Scheduling computer time	58%
Insufficient peripherals	57%
Not enough copies of software	54%
Insufficient teacher time	54%
Not enough simultaneous access to the Web	53%
Not enough supervision staff	52%
Lack of technical assistance	51%
Outdated local school network	49%
Not enough training opportunities	43%

were an insufficient number of computers and a lack of knowledge and/or skills on the part of the teachers. Many of the difficulties pointed out by respondents were related to a lack of funding for the necessary facilities, as shown by "insufficient number of computers", "insufficient peripherals", "not enough copies of software" and "outdated local school network". Other problems related to not having enough time or skills to be able to achieve the desired objectives using ICT.

Comparing Pelgrum's (2001) results with an earlier survey by Levy (1997) of CALL practitioners at higher education institutions or in private language schools in 1991, one can see a significant overlap, as shown in Table 7.3. It is significant that although Levy's survey was administered 10 years earlier than Pelgrum's, and with a specific focus on using technology for language learning in tertiary institutions and the private sector, the problems perceived by those involved with instructional technology remained largely unchanged.

Notably, the results of this survey also showed that lack of funding was one of the major difficulties faced by teachers and individuals who were looking to introduce technology into the classroom, as was a lack of time and insufficient training and expertise. The costs involved in using technology in the language classroom—or, indeed, in any other educational setting—extend well beyond the initial outlay of the hardware and software. In

TABLE 7.3

Blocks to Successful CALL Development ($n = 100$; from Levy, 1997)

Block	Percentage
Lack of time	35%
Lack of funding	24%
Lack of teacher training	10%
Lack of reward/recognition	9%
Teaching staff perceptions of CALL	8%
Level of expertise required	8%
Lack of agreed-on standards	7%
Level of acceptance in wider context	7%
Publishing problems	7%
Problems organizing teamwork	6%
Professional intransigence	6%
Lack of information sharing	5%
Lack of suitable software	5%
Other	13%

addition to this are the costs for supporting the technology and for training staff in using the facilities, as well as the costs in terms of time for development of materials. Even when the hardware and software are in place, development of CALL materials requires significant time and sufficient skills; before embarking on any type of course that utilizes CALL, teachers and administrators need to ensure that both of these requirements are met. Taking into consideration how much can be done in a limited period of time might mean that the scope of a project may need to be reduced, at least in the early stages. Similarly, very few sophisticated programs used for language teaching reached their complete level of sophistication in the early prototypes. It then becomes the responsibility of teachers and administrators who have made the decision to use CALL as a part of the language-learning environment to decide how much time to devote to development of materials and what level of sophistication for which to aim.

Other problems that can be encountered are of a more local nature, depending on the specific environment, even when all of the necessary hardware and software is already in place. For example, there are sometimes differences in what is required or expected by learners and CALL practitioners and what is made available to them by the administration. An example of this was presented in a study by Shimatani and Stockwell (2003), who surveyed university students about their preferred CALL usage times. Learners indicated that they would prefer to use CALL facilities in the evenings or on weekends; however, due to security concerns, the administration only allowed access to the facilities in the mornings before classes. This type of disparity is not uncommon, where the management is, quite understandably, concerned with the logistical aspects of maintaining the technology, and the immediate concerns of the learners are not as obvious as they might be to the teachers who are in direct contact with them on an everyday basis.

Difficulties also arise when trying to bring together two groups of students, in different institutions constrained by different curricula, to work toward a common goal. Apart from the logistical difficulties just outlined, there are also other fundamental hurdles, such as the differences in the learning objectives of the students at the institutions involved in the project. A means of catering for this is through an approach called "situated curricula" (Ligorio, Talamo, & Cesareni, 1999, cited in Ligorio, 2001). Originally a concept used in management literature to understand how people learn within organizations (Gherardi, Nicolini, & Odella, 1998), the situated curricula approach requires the "project manager" (or, in the case of a classroom situation, the teacher or course coordinator at each institution) to guide the participants in the design of the activities required of them while working within the constraints and available resources of the particular classroom. Each of the different environments involved in a project will naturally have its own individual constraints in terms of resources, facilities, and general learning objectives, and these must first be defined in order to determine the roles that the participants can play in collaborative projects.

An example of this was explained in a study by Gray and Stockwell (1998) of students in Australia and Japan interacting through e-mail. In this study, Australian learners were studying Japanese language, and the Japanese students were studying intercultural communication; hence, the topics of the interactions were centered on intercultural issues regarding Australia and Japan, but the language of the interactions was Japanese. The Australian students were assessed on the quantity and quality of the e-mails they produced, whereas the Japanese students were required to write reports based on the content of the interactions from an intercultural perspective. There was a marked gap in the resources at both of the institutions, which meant that there was also a difference in the way in which the project was implemented at a classroom level. Through finding the common ground between the two student groups, students were able to come together to work on a collaborative project, despite the fact that the requirements of the students varied in accordance with the curriculum at each institution.

Working internationally poses its own difficulties that need to be considered. As Fleming and Hiple (2004) argued, through videoconferencing, learners can be brought closer to interlocutors than in almost any other form of communication short of face-to-face interaction; yet, on the negative side, the interlocutors need to be online at the same time, which can be difficult if there are time differences. Other problems are faced even when learners interact through asynchronous forms of communication. Lehtonen and Tuomainen (2003), for example, described an international project that involved difficulties in timing of semesters because of the different systems in the countries of the participating students; the project needed to use the overlap period of several weeks as the window in which the projects could be carried out. Although the use of the overlap makes running international projects possible, care must be taken to ensure that this time is used effectively, particularly when one group must go the project at the beginning of a course semester and the other at the end. This has implications for the type and amount of training that can be given to learners before commencing the project in the early-starting group, and the kinds of posttask activities or assessments that can be given to the late-starting group.

One of the key factors when dealing with problems in a CALL environment is to be aware of the possible constraints. This means knowing what facilities, time, and funding are available, and working within the boundaries that these might impose. What separates CALL-based activities from more traditional classroom activities is that, due to the presence of technology, there is the potential for more things to go wrong. However, as has been a recurring theme throughout this chapter, forethought and preparation can help prevent many possible problems.

CONCLUSION

This chapter has dealt with the practice of CALL, specifically considering the types of CALL that can be used in the teaching of different language skills and language areas. A key point is that no single type of CALL is normally applicable to all language skills and areas—the range of skills and abilities required is simply too great. Conversely, a wide range of different technologies and materials may be brought to bear on one particular language skill or area, if it should be decided that it is to be a priority focus. CALL is multifaceted and encompasses a wide range of technologies, materials, and resources. The art of good practice comes out of choosing the right technologies, materials, and resources for the task at hand, both for the teacher and the student.

Finding and Keeping a Balance

One of the hardest things to achieve is a good balance between these various elements vying for the teacher's and the student's attention. The implementation of CALL can lean too heavily toward the technology, teacher-set objectives, or even the resources that are available at a given time. Levy (2006) commented that "effective CALL requires us to locate the optimal balance of approaches, resources and tools to meet the needs of particular learners in a particular learning context" (p. 1). This means that there is a need to constantly monitor the language-learning environment, and to evaluate whether the objectives are being met.

The introduction of the computer will invariably alter the classroom culture (Schofield, 1995), in terms of the interactions between the learners and the teacher and among the the learners themselves, as well as the types of tasks that are completed. The physical arrangement of the environment can and does change the ways in which students relate with one another and the teacher. Where students in traditional classrooms may sit at desks in one of any number of configurations, there is still an obvious difference when there are computers with monitors in front of every student. In the more conventional computer laboratory configurations, it is not uncommon to find students hiding behind computers, doing a little independent surfing on the Internet. Students have the potential to talk not only with the students beside or behind them, but also with any one of a virtually unlimited number of people who happen to be online at the same time, with minimal control by the teacher over what the students discuss or with whom and in what language.

Felix (2003b) argued that although there are aspects in which the technology excels, such as encouraging contact and communication among students and between students and the teacher, promoting active learning, and giving prompt feedback, there are still areas in which teaching through technology faces serious challenges. Perhaps one of the greatest characteristics of effective teachers—enthusiasm for the subject and for teaching—is often difficult to convey successfully through technology. In environments in which the predominant contact between teachers and learners is through the computer, the enthusiasm that may be visible through a teacher's facial expression or gestures is not as easily recognized. Ensuring this vital contact between the teacher and the learners is perhaps one of the greatest ways of overcoming this difficulty. Learners need the security of knowing that the teacher is available to support them rather than feeling that they have been abandoned to work on tasks independently.

A common trap for some CALL practitioners is to develop tasks that require learners to engage more with the technology itself than the language they are learning. Sometimes the technical complexity of a task can be underestimated. In such a situation, the teacher really needs to gain a sense of the students' technical skills and abilities before setting tasks that involve technology. The weighting in any assessment also has to be well judged and appropriate to indicate and convey priorities to students. For example, requiring learners to develop a Web page for which they are assessed on the sophistication of the page rather than on their target language usage could not be considered as contributing significantly to second-language development. To avoid this type of problem, the teacher must ensure that language-learning goals are set, and that the type of task selected contributes to realizing these goals.

Designing and using CALL materials effectively depends on setting clear learning goals, deciding which language skills and areas are to be targeted through CALL-based activities, and determining what types of activities are appropriate in achieving these goals. In addition, it is essential that CALL practitioners know the learners, their skills, and their backgrounds, and provide suitable training and guidance to help them understand exactly what is required of them.

Chapter 8

Technology

Making predictions about the future of state-of-the-art technologies is a hazardous business. Citing Microsoft founder Bill Gates, Fallows (2002) asserted that correct predictions about developments in technology will be regarded as obvious after the event, whereas incorrect predictions will be ridiculed. Developments in computers generally tend to fall short of or greatly exceed predictions made. Although there have been clear developments in the technology[1] used in CALL over the past 5 years that have affected the way in which CALL is used in the classroom, the impact of these changes in the future is difficult to determine. It is, however, possible to identify trends on the basis of existing and emerging technologies.

As technologies develop, they bring with them a range of functionalities that can be applied to language learning. Developments in smaller, more sophisticated storage devices—such as external USB optical and hard drives, memory sticks, flash memory, MP3s, and so forth—have led to much more flexibility, versatility, and access as far as language learning is concerned, in that it is far easier to store greater volumes of text, sound, and video. There has been much more development in storage than there has computer screens, for example, which have a lesser impact on language learning than storage devices might. It is to be expected, then, that new hardware and software are suited to different language-learning purposes. It is unlikely that a single advance in technology will be applicable to all the language skills that we may wish to teach, although resources such as language-learning Web sites are multifaceted and multifunctional, and are able to contain and support a wide range of language-learning activities.

---

[1]The word *technology*, as used in this chapter is applied not only to hardware, but also to developments in software.

Other technologies, such as the MP3 player, for instance, are more likely to be employed for a narrower purpose. Thus, technologies that are more suited to one skill than another will emerge at different times. Such differences raise the need for an examination of strengths and limitations of the new technologies to determine how these different technologies may best be used for specific language-learning goals.

When we look back at the technologies that have appeared over the last 20 years, we can see how the field of CALL has been affected, not only in practice but also in its discourse (e.g., Chapelle, 1990), research (e.g., Davies, 2001), and pedagogies (e.g., Felix, 2003b). The way in which language classes are taught has been dramatically affected by the technology. Early CALL consisted of learners working on simple tasks assigned by the teacher individually or in small groups, and words such as *collaborative learning* and *chat* held rather different meanings from what they came to mean after the advent of the Internet. In addition, before the Internet, teachers did not need to know how e-mail, chat, and audio- and videoconferencing contrasted with face-to-face teaching, telephone communication, and so on. Thus, the development of new technologies gives rise to completely new ways of thinking and communicating, and areas of research that could not have existed without such technologies. Furthermore, the introduction of new technologies may bring a complete reassessment or reappraisal of the pedagogy used. In some cases, this may lead to developing a completely new pedagogy, whereas in other cases all that is necessary is a readjustment of existing pedagogies.

The chapter provides an overview of some of the new technologies for language learning that have emerged in recent years. This is followed by a discussion of what CALL practitioners need to consider when adopting new technologies, and the choices they face. The chapter then continues with a discussion of the use of the new technologies outlined here, considering their potential applications and shortcomings, as well as examples from the literature describing how they have been used in language-learning contexts.

DESCRIPTION

New technologies for language learning span a wide range of areas, so a discussion of every new development in technology that can be used in language-learning contexts is just not possible. This chapter includes those items that have attracted the most attention in the literature and have appeared to have the most impact in CALL practice and research (see also Mills, D., 1999). The technologies described here are authoring software, learning management systems (LMS), and new developments that allow for hybridization of these tools; audio- and videoconferencing; artificial intelligence (AI) and intelligent systems; speech recognition and pronuncia-tion-training technologies; and mobile technologies. These descriptions provide a brief overview of the major issues associated with each of the de-velopments in technology, and references to other sources for more de-tailed information.

Authoring Software, LMS, and Hybridization

Authoring tools allow teachers to tailor activities to suit specific learning goals and objectives. They are varied in type, ranging from software for pro-ducing individual tasks to integrated systems that can be used to manage a large portion of a course. One of the better-known authoring programs is *Hot Potatoes*, developed by Half Baked Software. *Hot Potatoes* has gone through several stages of development, and now allows for processing of several languages other than English or other European languages, espe-cially those that incorporate 2-bit character sets (used with script languages such as Japanese or Chinese) rather than the typical 1-bit. *Hot Potatoes* is particularly noted for its flexibility, ease of authoring, and lightweight HTML files, which can be used by teachers in a range of environments. It of-fers a free license for teachers who register, provided that they are not using it for personal profit and that they make their activities available openly. Another authoring program is *MALTED* (multimedia authoring for lan-guage tutors and educational development), which aims to provide more of a focus on the multimedia aspects rather than pure textual feedback (see Bangs, 2003, for a more detailed description). Commercially available course tools are also commonly used, and many institutions have adopted one particular system to be used across a range of subject areas, not only sec-ond-language learning. Two of the more commonly used types of LMS—also sometimes referred to as *course tools, course management systems* (CMS), and *virtual learning environments* (VLE)—are *WebCT* and *BlackBoard*. Both are powerful tools, providing a bulletin board system (BBS), chat facilities, and e-mail links, as well as online activities, quizzes, and even submission of assignments electronically. In addition to these tools, there are also other commercially available authoring programs, many of them quite sophisti-cated. One such example is *Authorware*, which allows for quite complex inte-gration of multimedia and recording of results, including transferring the data into a LMS.

As Arneil and Holmes (2003) described, the types of online activities pro-vided through different authoring and LMS tools can be either client based (i.e., the test and all of the information is downloaded onto the user's com-puter, and all correction and marking occurs on the user's computer), or server based (i.e., only the necessary information is sent to the user's com-puter, and all corrections and marking occur on the server). *Hot Potatoes* and *MALTED* are examples of client-side tools, and *WebCT* and *BlackBoard* are examples of server-side tools.

Although there are many fundamental technical differences, one of the primary practical differences between individual authoring tools such as *Hot Potatoes* and course tools such as *BlackBoard* is the integration of activities into a whole, particularly for those who wish to record and manage grades in a comprehensive and systematic way. *Hot Potatoes*, for example, provides immediate feedback and scoring to the student, but it is difficult to have these scores made available to the teacher. In its standard form, it provides a facility for some CGI[2] tasks to be performed; most commonly, the ability to have results sent to the teacher by e-mail. When commencing an activity, students are faced with a pop-up screen where they enter their names and/or student numbers, and then are able to begin. Although useful, there are limitations with this format. The first is that many students make mistakes in entering their details, either through omission or error, and teachers receive an e-mail including incomplete information. Another is that teachers are faced with a large number of e-mails from the students that they must then methodically sort to be able to find the scores for each of the students. Any problems with the students' scores are seen only after all the e-mails are received, which may be some time after the event. Fortunately, the scripting involved in *Hot Potatoes* is relatively simple, and there are ways in which teachers can alter the script to forward results directly into a database such as *MySQL* (Daniels, 2004). This simplicity does, however, open up other avenues for problems. Normal classroom use is less of an issue; however, when used for testing, there are problems with security and cheating. For client-sided tools such as *Hot Potatoes*, the increasing technological skills of students means that, without strict supervision, many are able to look at the source code of the test being administered and find the correct answers. The later versions of *Hot Potatoes* allow teachers to store test information in a separate JavaScript file to try to make it a little more difficult for learners to break into, but unfortunately, even as the developers themselves admit, it is still relatively easy to access if students have computing knowledge.

A LMS, on the other hand, is far more secure and cannot be accessed without a login and password. Assessment can be carried out using these systems, and similar types of tasks can be designed as in *Hot Potatoes*, but any feedback is provided at the end of the activity when the results are submitted to the server (see Arneil & Holmes, 2003, for a detailed explanation). Scores for these activities can be automatically fed into a database where they can be made available to teachers. Despite these advantages, there are

---

[2]CGI stands for *common gateway interface*, and is used as a means of sending and processing information from static pages such as html.

also some problems associated with LMS. One of the major issues is that the costs involved can be prohibitive. As such, an alternative that has started to attract attention from a growing number of CALL materials developers recently has been open source (freeware) LMS, such as *Moodle* and *XOOPS* (Awaji, 2004). Both of these are PHP[3]-based, data-driven applications that allow much of the functionality of commercial sites.

A major problem with many types of LMS is that they have a reputation for lacking flexibility (see chap. 2 for more discussion of this point), and this can prove limiting in some ways. In order to overcome this lack of freedom, some teachers use combinations of existing authoring tools such as *Hot Potatoes* with their own systems (e.g., Shawback & Terhune, 2002). Although there is freedom from the point of view of what teachers are able to provide through such individually developed systems, one of the biggest limitations is the development time and the skills required for development. In addition, most LMSs operate using varied presentation formats, and they also handle data differently. Most did not, until very recently, allow for exchange of information between systems, or communication with authoring tools like *Hot Potatoes*, without additional adaptation.[4]

Recent developments have started to discuss extended functionality of these course tools, and allow them to be combined with other technologies, in what is termed as *hybridization*. More notable developments have included *SCORM* (sharable content object reference model) (Godwin-Jones, 2004) and *XML* (extensible markup language; Mishan & Strunz, 2003). *SCORM* is a means of sharing content between platforms and LMS (Godwin-Jones, 2004). Given that many of the current LMSs do not allow for exporting or importing of data across platforms, *SCORM* provides the means by which to incorporate a range of technologies by acting as the link between them. This ability to transfer data across different platforms and software boundaries allows CALL practitioners far more range in the artifacts that they are able to use, without being confined by policy decisions to adopt one format in preference to another. Not all practitioners agree that *SCORM* is the way to go with improving cross-tool compliance. Arneil and Holmes (2003) argued, for example, that *SCORM* is a "moving target" and that *XML* technologies are preferable due to their inclusion of metadata that makes them easier to find, use, catalogue, and assert the author's intellectual property. *XML* is a markup language that is used to describe the structure of data in meaningful ways, and assists in the input, output, and

---

[3]PHP is a freely available programming language used for Web-based applications. It allows easy integration with database applications such as *MySQL*. The acronym PHP originally stood for *personal home page*, but with increased sophistication that appeared in later versions, this was changed to *hypertext pre-processor*.

[4]A module for *Moodle* was recently developed by Bateson (2005) that allows *Hot Potatoes* activities to be easily incorporated into the *Moodle* LMS.

transmission of data (Hunter et al., 2004). The role of *XML* and *SCORM* in the language classroom is one of giving freedom to CALL practitioners. Rather than being limited to one platform such as *WebCT* or *Moodle*, these tools allow use of multiple platforms with transfer of data between them. These technologies also allow for integration of other types of data-base-driven applications as well, such as reference sources and learner profiling (see Beaudoin, 2004, for a detailed discussion of the educational uses of databases in CALL).

Learning management systems provide an environment in which a number of the tools that teachers require to manage a course can be located in one environment. With the increased ability to pass information from authoring software such as *Hot Potatoes* and *Authorware* to and from many LMSs using emerging technologies, such as *SCORM* and *XML*, teachers can choose the tools that they want and tailor these tools to their individual learning contexts. With greater choice comes greater diversity, and this diversity opens up the avenues for further development of ideas and concepts in the future.

Conferencing

One development in language learning that has been attracting increasing attention over the past few years is audio- and videoconferencing. Communicating with native speakers or even other students who are separated by large distances has long been described as an advantage of Internet technologies; however, with the exception of those institutions with large equipment and operating budgets, most interactions have been limited to text-based means such as e-mail, chat, and MOOs.

An important innovation affecting conferencing has been the spread of broadband Internet in recent years, which allows for transmission of very large amounts of information in a comparatively short amount of time. One such technology is DSL (digital subscriber line), which uses a standard telephone line to send and receive data at a rate of as much as 40 mbps, markedly faster than modems that allow a maximum of 56 kbps. In addition to DSL, fiberoptic and cable networks are also becoming more available to the general public, and although they can be even faster than DSL, they usually require some installation and setup to be used. Depending on the context, DSL, fiberoptic or cable networks may be all that is needed for Internet access needs, particularly in smaller schools or businesses. Most universities and larger businesses, on the other hand, have a T3 (also known as a DS-3) line in place, which is a high-speed connection that can transmit data at a rate of up to 45 mbps. With the greater speed and stability of connection of these technologies compared with modems, it is only a matter of time before modem connections become phased out completely. In some countries, such as Japan and South Korea, modems have been all but replaced by these technologies, and most people use DSL or, more recently, wireless broadband connections. Wireless technologies do have advantages for institutions and businesses in that they eliminate the need for putting cables and ports into rooms.

The increase in high-speed connections and browser-based conferencing technologies, and the decrease in cost of necessary equipment such as microphones and Web cameras, have made audio- and videoconferencing far more accessible than they were in the past, and many people now make personal use of them through standard desktop and even laptop computers. Some operating systems already come complete with conferencing software for both audio- and videoconferencing, such as Windows *NetMeeting* (see chap. 4 for more discussion on conferencing software). These advances in technology have brought about a number of studies into both audio- and videoconferencing, most notably in the field of distance education and learning (e.g., Hampel & Hauck, 2004; Strambi & Bouvet, 2003; Wang, 2004a, 2004b), although there have also been studies into environments that seek to give learners speaking opportunities outside of normal class hours (e.g., Levy & Kennedy, 2004). Through provision of audio channels available in conferencing, it is possible for learners to feel a sense of presence that is often missing in nonaudio means (see chap. 4 on CMC for further discussion). In addition, there is the capacity to correct language errors or pronunciation immediately.

Conferencing does not necessarily mean only the exchange of voice and images of the speakers. Hampel and Baber (2003), for example, described software developed at Open University entitled *Lyceum*, which enables a text chat, a whiteboard, space for concept mapping, a shared word processor, subgrouping "rooms" for students to discuss matters in smaller groups, file sharing, and file transfer (see the earlier discussion of *Lyceum* in chap. 2). Similar functionality is also available in *NetMeeting*, with desktop sharing allowing for shared control of Web browsers and word processors (Levy & Kennedy, 2004).

At present, there are still problems with bandwidth that need to be overcome. Despite the speeds that have been achieved through DSL, cable, or fiberoptic networks, many CALL users still comment on the problems with clarity of sound or use of video imaging, and the effect of using video on overall computer performance (e.g., Hampel & Baber, 2003). Use of audio alone can help to ease this problem, but even the current quality of sound production is equivalent or slightly lower in standard than it is with a telephone. When used for teaching groups of students, other difficulties arise when more than one person speaks at the same time, making it difficult for teachers to differentiate individual voices.

Collaborative learning that can be achieved through conferencing is supported by socioconstructivist approaches to language learning (e.g., Felix, 2002). Thus, conferencing offers good potential as a tool in language learning, and the fact that learners can orally interact with one another and the teacher despite being separated geographically is motivating for all involved.

Artificial Intelligence (AI)

Artificial intelligence is an often misunderstood field, surrounded in its early days by overly extravagant claims, many of which were not achieved by the technologies of the time (Levy, 1997). As Dusquette and Barrière (2001) pointed out, much of the AI research of the 1980s was full of promises that were not realized, and the resultant decrease in funding brought with it a revision of the expectations, a subdivision of the problems, and a narrowing of the focus. AI resources developed for language learning have generally been termed as ICALL, or intelligent computer assisted language learning, which promised to have a significant impact in the future (Kohn, 1994). Although more than a decade has passed since such claims were made, the contribution of ICALL thus far has been, as with other forms of AI, somewhat modest. On the other hand, recent work by such CALL developers as Dodigovic (2005), Heift (2002, 2003) and Heift and Schulze (2003) demonstrate that real progress is being achieved in this area.

The acronym ICALL itself gives little indication of what it actually represents (Schulze, 2001). In the early 1990s, Matthews (1993) predicted that "the obvious AI research areas from which ICALL should be able to draw the most insights are Natural Language Processing (NLP) and Intelligent Tutoring Systems (ITSs)" (p. 6). A survey of the literature shows that the majority of research in the field in recent years does indeed fit into these two areas. NLP is "concerned with the computational modelling of the perception, analysis, interpretation, and generation of natural language" (Jager, 2001, p. 102). NLP systems work by parsing text for specific features, and responding to certain keywords contained within the text. NLP has faced several challenges in its development, not the least of which is the ambiguity of human languages (see Farghaly, 2003, for a detailed discussion). Early NLP was concerned with machine translation (MT) and story comprehension, both of which are difficult tasks that require access to sufficient background knowledge to contextualize the input received (Dusquette & Barrière, 2001). Although there is continuing research in the field of MT (e.g., Wilks, 2004), of more interest to ICALL has been the ability to parse learner input for errors.

It is perhaps for this reason that it not unusual to see NLP exist as a component within intelligent tutoring systems (ITSs), the other major area of

ICALL (e.g., Heift & Nicholson, 2001). An ITS designed especially for language learning is an ILTS (intelligent language tutoring system), which is a computer-based instructional system designed to reproduce the behavior of a human tutor in its ability to adapt to the learning needs of individual students (Moundridou & Virvou, 2003). As Kang and Maciejewski (2000) explained, although there are variations in the architecture of ILTSs, they will generally consist of an expert knowledge module (providing the information to be taught), the student model module (the dynamic representation of a student's competence), a tutoring module (the component that designs and regulates instructional interactions with the student), and the user interface model (controlling the interactions between the system and the student). When coupled with NLP they have the ability to parse learner input, locate errors, and provide individualized feedback in response to these errors.

Although there is potential for the use of NLP in language learning, the costs involved in developing such systems are high and as such should only be considered if other, more conventional tools are not available (see Jager, 2001, for a detailed overview of the use of NLP in language learning). ILTSs are typically very complex systems—not only in terms of their construct, but also in their interface—meaning that teachers have not been able to build their own expertise into the systems without the help of knowledge engineers (Tokuda & Chen, 2004). Thus, although the argument for AI—and more specifically ILTSs and NLP—providing an interface that can patiently deal with learner errors is a strong one, the reality of a technology that can deal with errors on a paragraph level rather than a sentence level is still some way off.

Speech Recognition and Pronunciation-Training Technology

The significance of the development of speech technologies may be indicated by the fact that in 1999 an entire issue of *CALICO* was devoted to speech technologies for language learning; it contained 10 articles, some of which are described here. The fundamental technologies for automated speech recognition (ASR) and pronunciation training share a number of features; hence, they are dealt with together in this section. Most significantly, both must identify sounds and code them into a digital format that can be compared against models stored in the expert knowledge module. There are also differences that distinguish the two. Where ASR systems will generally refer the input to a language parser for processing, pronunciation software will generally seek discrete points of pronunciation within the input for the purpose of correcting it.

ASR technology consists of two main forms: discrete and continuous (Harless et al., 1999). Discrete speech recognition requires the developers to identify all possible expected utterances and to store these for comparison for the individual exercise. This requires the utterances to be short, and there are usually only a small number of possible responses that are compared against a set of speech models or templates, as in multiple-choice-type questions. Continuous speech recognition, on the other hand, is more complex, and allows the system to recognize anything spoken in the language, thus allowing for open-ended questions. Continuous ASR systems generally work using the hidden Markov model (HMM), which consists of two major components: inventories of statistical models that represent given phonemes in their phonological context, and word lists with phonemic translations from which vocabulary can be recognized (Precoda, 2004). Once the vocabulary items are recognized, they are pieced together into strings that can then be processed by parsing software.

Applications of ASR are diverse. One such usage is in an ILTS, as described earlier. Learners respond to cues given them by the computer, and their digitized oral output is parsed, errors are located, and feedback is provided in response to these errors. An example from Mostow and Aist (2001) illustrates how ASR can be used to help with reading in young children learning English: Learners read aloud passages provided for them on the screen, and the computer "listens" to the output and corrects errors as they occur. Other uses are very creative. Harless et al. (1999) wrote of multimedia software in which learners are required to "interrogate" their digital interlocutors in Arabic by asking guided questions that prompt videoed responses from the computer. A similar interface was used by Bernstein et al. (1999) in their study, in which learners interacted in Japanese to complete preset language functions.

The other major use of speech technologies is pronunciation training. The applications of pronunciation training software are also varied, and may look at certain aspects of learner output, such as segmentals (i.e., individual sounds) or suprasegmentals (prosodic features such as stress and intonation). Most software relies on visual representations of the sounds produced by learners compared with the correct model (e.g., Chun, 1998; Dalby & Kewley-Port, 1999; Hardison, 2004). Many of the projects that are currently being conducted in pronunciation training are based on software that has been developed by project groups at the institutions themselves, but there are also studies that adapt tools that are often used in speech pathology and voice training for musicians, such as *Sona-Match* by Kay Elemetrics, to pronunciation training in second-language education (e.g., Carey, 2004).

Speech technologies face a number of challenges as well. As Precoda (2004) commented, due to differences in size and shape of the vocal tract and individual speech styles, creating a speaker-independent recognizer that will operate effectively for any language learner requires a large speaker population to provide sufficient acoustic data on which to base the system. Although in mainstream languages these large bodies may be available, there are potential problems for less commonly taught languages. A further problem is that of accuracy. Some inaccuracies are caused by individual differences, whereas others are the result of improper microphone positioning (Mostow & Aist, 1999). Researchers are, however, always looking for ways to improve the accuracy of speech recognition software, such as using semantic analysis (Erdogan, Sarikaya, Chen, Gao, & Picheny, 2005), or developing new models, like the extended union model that seeks to solve problems such as short temporal corruption (Chan & Siu, 2005).

Developments in speech technology enable new interfaces previously not possible. As Vila, Lim, and Anajpure (2004) noted, emerging technologies such as *VoiceXML* widen the possibilities of using the Internet with voice-response applications. Through these technologies, teachers have the ability to link Web-based voice input to server-based ASR and pronunciation training applications, expanding opportunities for self-study as well as links with other Internet-based technologies, such as LMS or other authoring tools.

Mobile Learning

An area that has seen an increasing amount of attention in the past few years has been mobile learning (e.g., Kukulska-Hulme & Traxler, 2005). This kind of learning is associated with mobile phones, handheld PDAs, and even small notebook computers connected to wireless networks. The rationale given for the use of mobile technologies is that they enable a transition from the occasional, supplemental use associated with computer labs, to frequent and integral use (Roschelle, 2003, p. 260).

Mobile phone technologies are not, of course, just limited to email and SMS, although at this stage, this is by far the most common usage in language learning (see chapter 4 on CMC for a description of email for language learning). These types of projects may include situations in which students exchange e-mails with one another or the teacher (e.g., Aizawa & Kiernan, 2003), or receive mini-lessons through their mobile phones (Thornton & Houser, 2001). As mobile phones have developed to a point of which they are able to access the Internet, the possibilities to include other types of activities have also increased. Taylor and Gitsaki (2003), for example, discussed a study in which learners were required to use their mobile phones to search for information in English language pages. Many phones are now capable of handling *Flash MX* applications, which allows for presentation of information and interactive interfaces of a rather complex nature, some of which allow handling of results of tasks to databases contained within current LMS (Houser & Thornton, 2004). In many countries, mobile phones have had the capability to view video files for some years now, but the costs involved in downloading video segments, even those that are only a few seconds long, make their uses limited in most language-learning environments. Still, with better compression formats, including *Flash MX*, the size of the video files may be greatly reduced, which proportionately will reduce the costs of viewing the videos.

One consideration here is that the nature of many mobile phone activities are similar to those of a normal desktop computer, but with the small screens and inconvenient keypads (Thornton & Houser, 2002) one might expect there to be little motivation for the user to move to a new technology. However, from the point of view of horizontal integration (discussed in chap. 2), the widespread acceptance of mobile phones in the wider community means that their potential for language learning should at least be seriously considered. A novel idea of integrating the technology was offered by Price and Rogers (2004), who suggested that use of mobile technologies can allow learners to interact with the physical world while receiving digital information through mobile technologies. In other words, the learners can interact with each other or use specifically designed CALL materials while in an actual environment, such as a shopping centre or park. In this way, learners can perform collaborative activities in the real world while communicating through tools such as chat or e-mail, providing the potential for a more authentic environment for learners.

The examples provided here are only a sample of the increasing range of technologies appearing for language learning. However, they are, we believe, representative of the current trends.

DISCUSSION

As Levy (1997) noted, CALL at any given point is often a reflection of the technology that exists at that time. When a new technology is made available, it is not uncommon for there to be a surge towards that new technology as it appears, often followed by a number of papers and projects utilizing it. Some of the trends stay, but many of them disappear after a brief appearance in the limelight. This leads us to ask the question of why CALL practitioners constantly seek new technological alternatives. The answer to this may come in part from an interesting observation made by Bax (2003). He argued that a fallacy exists in many CALL practitioners' attitudes toward technologies, which maintain that any new technology (including software) should be able to do everything, and that if it had more features it would be inherently more effective. In other words, there is an underlying belief that with the release of a new technology, the solution to many of the problems faced in CALL has arrived. Taking this belief one step further, it can be extrapolated to mean that there is a common belief that technology has the potential to solve problems of pedagogy.

This has serious implications for the role of technology in CALL, and as such raises several key questions: What are the factors that underlie our choices of technology? Are our decisions regarding choice of new technologies founded on the sophistication of the new technology rather than on solid pedagogical foundation? What kind of pedagogical benefits do these new technologies afford us? What kinds of considerations do the introductions of new technologies carry with them? These questions cause us to consider the impact of the technology, in terms of the choices that we as CALL practitioners are faced with, at both administrative and classroom levels, and the challenges that these choices bring.

Making Choices of Technology

This is a time of choice for CALL practitioners. The range of CALL software that is currently available—including various courseware, authoring, and communication tools—is greater than it has ever been. There are a number of well-conceived and well-executed examples of CALL projects with clearly defined learning goals and pedagogies described at various points in this book. However, a good deal of what happens in the classroom can be technology driven if language teachers are not careful, and is founded on teachers attempting to fit learning goals into what the technology permits. Considering that CALL as a discipline is one that is dependent on technology, it is not altogether surprising that practice in the classroom is often influenced by new technological advances. Nelson and Oliver (1999), however, recommended caution in implementing technology in the classroom:

> We should not be using a computer just because we can and we should not be using more advanced (i.e., complicated) technologies when something simpler will do. Each technological advancement (since the paper and pen) creates different venues for learning, and supports different learning and teaching styles. The computer may or may not be better than teaching on a black-board; it all depends on what the teacher wants students to learn and how they plan to teach. (p. 101)

We must take care to ensure that our decision to use a particular technology is because it satisfies the pedagogical needs of the activity at hand, rather than for other reasons, such as the current trends in the use of a technology, or external forces such as pressure from the institution to use a new technology because of the considerable funds that the institution has invested into that technology. It is perhaps this latter factor that holds the most significance for many teachers associated with CALL. Sadly, the end users of technology (i.e., the teachers) are often not involved in the decision-making process at all. Cuban (2001) described the process behind the introduction of a multimillion-dollar state-of-the-art learning center, as it was termed at the time, at Stanford University in the 1960s. The swift decay of the center, according to Cuban, was largely attributed to the fact that few of the teaching staff were involved in the design of the facilities, technical support was only funded for a short time, breakdowns were commonplace, and the facilities were superseded by newer machines shortly after they were built. Sometimes these decisions are made by institutions for fiscal reasons. Little (2001) posited that the mentality behind introducing self-access CALL centers is "as a way of saving money: buy machines and learning materials now and they will serve you for five years; hire teachers now and you must pay them for as long as you offer the courses they staff" (p. 29). Thus, the choices of technology are very often made at an institutional level, and teachers are left to decide how best to use what has been provided for them (see chap. 2 for a detailed discussion of technology integration at an institutional level).

There are, however, other factors that affect the selection of a given technology outside that of the institution. There are times when the language teachers themselves are the ones who are given the choice of what technology to introduce, often without the expertise to make an informed decision alone. In such cases, it is common for these teachers to seek the advice of the people around them who have their own experiences with using technology. Given this, it is perhaps not surprising that the usage of CALL often seems to take on its own individual "flavor" depending on the country or region. Although one of the underpinning features of CALL is that it enhances globalization and international exchange of information, the practice of CALL itself tends to be localized, and the information and advice about which technologies to use and how they may be used often come from the people around us.

This localization of CALL could be attributed to the fact that developments in technology are often specific to a country/region. With the exception of numerous generic technologies, such as the latest microprocessor or newest data storage device, many new technologies are locally distributed. In countries where the technology is more readily available, it follows that much of the talk of CALL is more likely to be affected by these changes in technology. A simple example of this is the mobile phone. Mobile phones in one region are generally not transferable to others, and are highly dependent on the telecommunications network provider. Internet-capable mobile phones are appearing in many regions of the world now, but it follows that the regions in which they first appear will also be ahead in research in a given field. Much of the current research that appears in the international literature in the field of mobile phone technology for second-language learning comes from Japan. Given that more than 95% of the student population in Japan own mobile phones (Thornton & Houser, 2002), it is natural that CALL practitioners see the existence of a technology that has the potential to be applied to a language-learning environment.

As described earlier, in countries like Japan and South Korea, modems have been virtually phased out altogether, and DSL, cable, fiberoptic, or wireless technologies are the only options available to consumers. It thus follows that considerations of bandwidth for streaming video are not really a major issue. In fact, many television stations in South Korea make their popular programs available to consumers over the Internet at little or no cost. The picture quality is quite reasonable and the speed of streaming means that there is very little buffering, providing a very smooth picture. In countries where broadband Internet technologies are not as readily available, services that rely on a lot of bandwidth, such as video, tend to be greatly reduced. In contrast, at the time of writing this book, many people in the United Kingdom still rely on modems for their Internet access. In response to this, Open University *Lyceum* is designed to provide satisfactory quality of audio even through a 56K modem (Rosell-Aguilar, 2004).

There are many reasons to choose a particular technology for use in language-learning environments, some of them prompted at an institutional level, and some of them due to local conditions and knowledge. As a general observation, there are six main points that must be kept in mind when making choices of new technologies. Note that the points listed here do not replace those outlined in chapter 3, but should be considered as a first step when choosing a technology for a language-learning context.

1. Hardware and software requirements of the new technology.
2. Compatibility with other existing technologies.
3. Institutional support of the technology.
4. Ease of learning how to use the technology.
5. The general scope of use of the new technology.
6. Expected lifespan of the new technology.

First, the use of any new technology is generally dependent on the existing technologies, such as the make and capacity of computers in a computer laboratory. Because many new technologies will often require a large amount of processing power and memory, there is a need to ensure that the minimum requirements are in place to use these technologies, including having the latest operating systems or Internet browsers. Other, lesser-known problems that arise are caused by not having suitable ports on the machines (i.e., sufficient numbers of USB ports, FireWire ports, etc.) or the ability to work over network firewalls. Second, many new technologies are not compatible with the software and hardware already running on computers. Software and hardware manufacturers will generally clearly mark on their packaging if there is a known incompatibility, but failure to confirm can be an expensive and time-consuming exercise, particularly if there are large numbers of machines.

The third point—the issue of institutional support—is one that has been raised at different times throughout this book, and is a key factor in determining what technologies should be used and how fully they will be implemented. With institutional support, there is a greater move by staff to embrace new technologies than when decisions are left up to an interested few (e.g., Gillespie & Barr, 2002). Without the support of the institution, there is greater burden and responsibility on teachers to understand a new technology, and even to deal with technical problems alone.

The next point regarding ease of learning a new technology relates to both the teacher who is responsible for facilitating classes with the technology, and the students who use the technology in class. As described in the previous chapter, independent studies by both Levy (1997) and Pelgrum (2001) showed that language teachers and administrators found lack of knowledge or skills to be one of the biggest barriers to adopting new technologies, a point supported by Debski and Gruba (1999), who argued that unfamiliarity with advanced technologies can lead to uncertainty toward using them. Both teachers and learners must be convinced that the time spent in learning how to use a new technology compares favorably against the possible benefits. Selection of certain types of authoring software, such as *Hot Potatoes* (described earlier in the chapter), is sometimes based on the fact that it is easy to develop tasks in the midst of increasing demands placed on teachers. As such, if a new technology requires too much time to learn how to use, then it is likely to be abandoned in favor of easier alternatives.

The fifth point is concerned with how widely the technology is being used. The widespread use of a technology is often an indication of its stability and general acceptance, and has implications in terms of the support, the software, and peripherals available. The final point—the expected life span of a new technology—is one aspect that in many ways is difficult to judge. Computer hard disks, for example, have an expected life of 2 years with constant usage, but many last much longer, whereas others break down well short of this. Manufacturers of different types of hardware will often provide advice as to how long they expect their products to last. Software is harder to judge and is often dependent on developments in hardware, but observation of the general market for the software can be a good indicator.

Applying New Technologies to Language Learning

Authoring tools such as *Hot Potatoes* and LMSs such as *BlackBoard* can offer a range of useful activities such as cloze, multiple-choice, and matching exercises (e.g., Shawback & Terhune, 2002), forums through which to discuss language-learning issues, or a means to submit assignments (e.g., see Möllering, 2000). Combining various technologies to achieve specific language-learning goals also provides learners with diversity and can help to increase productivity and motivation, as has been the subject of countless studies. Failure to take appropriate measures to ensure that the technology is matched to the goals of learning, and that the management of this technology is properly addressed, can have adverse effects, as described by Dron (2003), who detailed how the technical and organisational problems caused by the blended delivery mode contributed directly to student anxiety and adversely affected learning.

Developments in technologies such as *XML* and *SCORM* have also extended the potential of these tools, in that they open up the exchange of information between formats and platforms. This has implications for language teaching in two ways. First, it means that teachers are not bound to one platform, by which they would have to make decisions from the outset to use an LMS such as *WebCT* or *Moodle*, and changing platforms would result in the loss of the materials stored within the LMS. Having the content of activities in a database format that can be accessed by different platforms means that teachers can change platforms without the need for rewriting content from scratch (Colpaert, 2004). Colpaert also asserted that the information in a single database can be shared by different applications, meaning that the same content can be used for presentation of new information in tutorial formats, and at the same time can form the bases of quizzes and tests. Second, the use of databases allows for knowledge pooling (e.g., Armitage & Bowerman, 2002; Cushion, 2004) and reusability (Ward, 2002). As the name suggests, through knowledge pooling, groups of teachers who are teaching similar content can put their resources together in a single location that can be accessed by everyone in the group. Over time, the pooled resources can be stored in such a way as to be easily retrieved by the teachers who are looking for content to suit a particular task or activity, so that they can then dedicate their time to the creation of new means and methods of teaching content without spending time on recreating content that already exists. Reusability, on the other hand, refers to a situation in which, once a language-learning technology has been developed, it is possible to be reused for different languages by simply replacing the required language-content data.

A good example of how *XML* technologies contribute to reusability was given by Ward (2002). She explained that through the development of a single processing engine, *XML* data files can be created for three different languages, in this case Nawat (a language from El Salvador), Akan (a Ghanian language), and Irish (Gaelic). The processing engine, written in *XSL*, was designed to look for certain predetermined tags in the specified

XML files that gave definitions of the data contained in it. *XML* allows for conversion between different formats, and as such allows for current materials to be converted to new formats, as well as making data accessible to different software applications, which helps to alleviate the problem described in chapter 2 of the limited lifespan of CALL materials. Another example of the use of *XML* for language learning was provided by Mishan and Strunz (2003). In their study, they related the development of an electronic resource book for authentic materials, called E-RAM. The objective of the study was to design a tool that could generate tasks that matched given authentic text. By attaching four levels of tags to a description of a language-learning task, Mishan and Strunz were able to progressively refine the description of the task, in terms of the genre (e.g., broadcast media), the discourse type (e.g., news item), the communicative purpose (e.g., provocative), and finally, the type of task (e.g., inferencing). Teachers were then able to browse through each of the four levels of menus to decide what type of task they wished to use.

Video- and audioconferencing technologies have contributed greatly to creating opportunities for language learners, either in addition to their current situation (e.g., Levy & Kennedy, 2004), or where such opportunities do not normally exist (e.g., Hampel & Hauck, 2004). Such technologies are not necessary in a second-language (as opposed to foreign-language) environment in which learners have ready access to native speakers outside of class time, and nor would they be suitable if there are sufficient options for learners to engage in face-to-face interactions with native speakers or even other learners. It is, however, necessary to consider the hardware and software that currently exist in the institution, and whether or not it is capable of handling the requirements for the conferencing technologies. It would also be pointless investing in the technologies if there are no speakers with whom the students can interact. Locating suitable partners who can be available at the same time as the students is not necessarily as easy as it seems, and it is better to ensure that these logistical issues are resolved before any financial layouts are made.

Levy and Kennedy (2004) provided an example of a study that utilized audioconferencing technologies in the teaching of Italian, in which the audioconferencing was used as a support for their ongoing independent learning to develop a balance between fluency and accuracy. In Levy and Kennedy's study, in addition to communicating through audioconferencing, learners interacted with each other and their teachers by e-mail, in person, and through the telephone in achieving their project goal of producing materials for the university Italian Studies Web site. A feature of the approach adopted by Levy and Kennedy was the sequence of tasks, ranging from a more protected environment through to an open one, designed to reduce stress and to familiarize the learners with the technology. Learners conversed through the audioconferencing software, in this case *NetMeeting*, initially with just the teacher, then with a fellow participant, next with a native-speaking stranger through the local university network, and finally with a native-speaking stranger in Italy or anywhere outside of the university. All of the conferencing sessions were recorded using video capture cards, a sound mixer, and a video recorder. After the conferencing sessions, learners participated in stimulated reflection sessions in which they considered both the language used during the conference and the process behind their conversations, while watching the computer screen contents and listening to their recorded interactions. This study demonstrates that the use of audioconferencing technologies need not be simply a technically sophisticated replication of what happens outside of the classroom, but rather a means by which the affordances of the technology allow for language-learning opportunities.

The developments in AI for language learning (ICALL), and in particular ILTSs, allow for a form of highly individualized learning that is difficult to achieve in many language-learning environments, particularly when there is a single teacher responsible for large numbers of students. Similarly, speech technologies can also be used to provide specific feedback for learners, both in terms of simple oral practice and for helping learners correct certain aspects of their oral output. In saying this, they are not without their limitations, not the least of which being the cost concerns. Unless institutions have large development budgets, or are able to purchase commercially available products, such technology is not really a practical option when compared to a human teacher.

An example of an ILTS was offered by Nagata (2002), who explained a system of parsing input from learners of Japanese called *BANZAI*, consisting of a lexicon, a morphological generator, a word segmenter, a morphological parser, a syntactic parser, an error detector, and a feedback generator. Through the software, the program identifies errors produced by learners in response to one of five different types of production-based tasks, ranging from word level, to phrase or clause level, sentence level, and through to paragraph level. The program then provides the learners with detailed feedback outlining each of the errors made and explanations of what is wrong. A feature of the system is that it does not require learners to simply read and respond, but rather it includes a listening component, while at the same time giving learners access to resources for vocabulary and grammar, as well as audio files of the correct answers.

Mobile technologies are one area on which a lot of attention has been focused lately, and there are those who argue that mobile technologies provide an alternative to the CALL laboratory (e.g., Kluge, 2002). The prospect of mobile technologies replacing desktop computers in a CALL laboratory is perhaps not realistic, and the problems associated with mobile technologies, such as the small size of the screen and keyboard described earlier, place certain restrictions on their widespread use as a total replacement for the computer. Although large numbers of students now own mobile phones (see Thornton & Houser, 2002, for an overview), expecting that all students will buy a PDA (personal digital assistant) may be somewhat unreasonable. This is not to say that mobile learning does not have a place in the language-learning curriculum, and there are studies that indicate positive reactions from students (e.g., Taylor & Gitsaki, 2003). There are important considerations that must be made, however, before expecting learners to utilize mobile technologies, including mobile phones. Even in countries such as Japan, where it is very common for students to have mobile phones, it is rare that number of students with phones actually reaches 100%. This means that those students who do not have the technology may be excluded from their use, or otherwise may feel obliged to buy one. In addition, using a mobile phone costs money. Even though the costs for sending and receiving messages or accessing the Internet may be very modest, it is still a cost that students themselves must bear, and it is possible that some may be opposed to it.

Innovative studies using mobile technologies are starting to appear in the literature. An interesting study was performed by Kiernan and Aizawa (2004), in which Japanese learners of English were required to perform tasks that involved exchange of information through mobile phone-based e-mail. In the first task, learners were paired up and one learner was given a series of pictures depicting an action or an event, while the other learner was given the same pictures but in jumbled order. Based on the information provided by the first learner, the jumbled pictures were put into the correct order. After the pictures were sorted, the second learner gave an account of the story according to the pictures. Finally, this was followed by a role-play activity based on the information included in the pictures. One of the aspects of this study was that it went beyond simple discussion of assigned topics, instead requiring learners to participate in an activity that meant that exchange of information was necessary. Due to the slow input of text—particularly in English—into the mobile phones, the authors suggested that mobile phone e-mail activities were more suited to lower-proficiency learners, because their limited language skills are less affected by the keypad.

New Technologies, New Problems

Although CALL has done much to shape the way in which we teach and learn languages, and has opened up many opportunities for learning and exchange that were not possible through more traditional means, there have also been some rather negative, unexpected outcomes as a result of the use of new technologies in the language classroom. Through the use of the Internet and communication tools, the computer offers learners a wide range of resources that were once difficult to access, but at the same time it has also made it easier to use these materials in ways that are not appropriate. The digital format of material accessible over the Internet can be copied and transferred very easily, giving rise to three major areas of digital cheating over the past few years:

1. Plagiarism and copying.
2. Hacking into online quizzes.
3. Machine translation.

There is no doubt that technology has made plagiarism much easier for learners than it once was. Where once students who plagiarized had to physically write out text from sources such as books and periodicals, the copy and paste function available in most available software makes it far simpler. Although copying materials written by other students on word processors has been possible for many years, a more recent avenue for plagiarism is the use of resources from the Internet (Barnes, 2003; Hopkins, 2004; Ryan, 1998). The Internet provides a huge amount of information in almost every field, in several different languages. Learners who are more interested in passing their subjects than in learning through them often yield to the temptation to write reports made up almost exclusively of information culled from the Internet in the target language. Other learners e-mail reports to classmates or friends and pass them off as their own. Fortunately, there are also tools being developed that can help to stem some of these problems. Decoo and Colpaert (2002), for example, described software that searches matching strings, identifying matches from the Internet to locate cases of plagiarism in student writing. Similarly, Stockwell and Nozawa (2004) developed a system that allowed teachers to check essays against all other essays in the student database to determine whether or not cloned assignments were submitted.

Hacking into the source code to view the answers to questions is perhaps an inevitable aspect of using online quizzes (Arneil & Holmes, 2003). As described earlier in the chapter, client-side activities such as *Hot Potatoes* embed the answers to the questions within the source code, making it possible for learners with sufficient computing knowledge to access this information. The problem can be alleviated to a degree through the use of server-side quizzes, but this has the potential to reduce the interactivity of the activity, because information must be sent back and forth to the server rather than accessing all of the information from one location, as is done in client-side applications. As Arneil and Holmes noted, rather than technical solutions, the ideal key to solving this form of cheating is to make sure that supervision of the quizzes is sufficient to discourage such behavior.

A final problem that has started to appear is the use of machine translation (e.g., Kempfert, 2002). When learners are required to submit essays in the target language, students with lower proficiency have been seen to write their essays in their L1, and then use translation software to translate it (Stockwell & Nozawa, 2004). There have been an increasing number of free online translation services by which learners can input text in one language and it is automatically and immediately translated. Although these machine translations are often unnatural, they are typically grammatically correct, and determining what has been written by the learner and what has been machine translated is sometimes difficult. As the technology develops, this problem will become increasingly difficult to distinguish, and teachers who aim at assessing writing skills may be faced with the choice of requiring learners to perform written tasks in a classroom environment using computers that do not have access to the Internet, or by using filtering software[5] so that these sites cannot be used. This does not, however, solve the issue of work that is done outside the classroom, and teachers will most certainly need to adopt creative measures to avoid this.

Cheating such as plagiarism is an ever-present problem, not just in language teaching, but in all fields of education. Eliminating it entirely is unlikely to happen, and curtailing it will require preventative measures such as education and counseling (Pecorari, 2003). Rather than seeking ways to combat the problems caused by the technology with technology, the most practical solution is to ensure that the learners themselves are made aware of the moral and ethical issues involved.

CONCLUSION

Too often decisions regarding the introduction of new technologies in schools, colleges, and universities are made at the administrative level without the input of the people who will be using them. There is a need for the institution and the end user to work together in the decision-making process, to develop a clear picture of what is needed, to understand what is practical, and to recognize the strategies and resources required to keep the technologies properly maintained. CALL relies heavily on integration at different levels to be successful, a topic we have discussed at points throughout this book. Integration is also the theme of the next chapter.

Key areas in which developments have been made are in technologies such as *XML* and *SCORM* that allow for hybrid learning platforms. As a re-

---

[5]Filtering software can block access to certain Web sites. Parents with young children often use these to stop their children from accessing inappropriate material over the Internet.

sult, teachers are afforded far more flexibility using these tools. In addition, these technologies allow the reusability of both content and of the systems themselves, which, in turn, leads to greater longevity of the CALL materials and resources that are created. Knowledge pooling is also possible, and teachers can work collaboratively in the development of a resource base that can be accessed by people developing new systems without the need for re-inventing the wheel. Conferencing technologies provide more opportunities for learners to use their language in oral interactions, but at the same time bring with them the responsibility of developing new pedagogies for enhancing learning. ICALL and automated speech recognition have shown great potential to help learners in the personalized nature of the feedback it can offer, but at this stage the technologies are limited due to their complexity. Still, with advances in system architecture and the human–computer interface making it easier for nontechnicians to use expert systems (Tokuda & Chen, 2004), and access to some online and open source programs over the Internet now available (see Coniam, 2004), they are becoming more accessible for language teachers. Finally, mobile technologies are attracting more attention, driven by the steady increase in the availability of mobile devices and their increased functionality. Freedom of when and where learners study is an attraction, but limitations such as keyboard and screen size will likely restrict the range of language-learning activities that can be undertaken.

Chapter 9

Integration

Although not regarded as one of the key dimensions in this book, integration certainly qualifies as an important topic. It has been a recurring theme: It was discussed at length in chapter 2, on design, and in the chapters 7 and 8, on practice and technology, respectively. The idea has provided a background for much of the discussion in the book, because integration encapsulates the idea of elements combining together to form an effective whole, and thereby it relates strongly to practical matters of implementation. In fact, integration has been a topic of discussion in CALL for many years. For example, Robinson (1991) reported on the conclusions of two research studies that highlighted "the importance of integrating individual CALL work with the total program of language instruction, including the classroom, rather than configuring it as an independent, supplementary activity" (p. 160). Hardisty and Windeatt (1989) emphasized precomputer and postcomputer work as well as work at the computer. They valued the importance of integration not only at the lesson level but also at the curriculum level. Hillier (1990) concluded that student training, teacher training, and class scheduling were the most important elements for integrating computer work into their program (see also Levy, 1993). Taken as a whole, it seems there are quite a number of elements in play when thinking about integration and CALL.

The question of integration really relates to the ways in which the various elements influencing the use of new technology in language learning are brought together and managed in order to create a successful CALL environment. The technology itself certainly adds another dimension. We need to understand more clearly what is involved in successful integration and how we might break down the concept into a number of practical ideas and strategies.

In the design chapter, the terms *vertical* and *horizontal integration* were introduced to emphasize the important issue of continuity. Such a perspective requires a greater appreciation of the student's technology-related learning experiences outside the language class—for example, in other courses within the institution or at home. It also requires more of an understanding of the wider technological infrastructure of the school or university, such as the favored hardware and software and the support and resources available. This general approach is broader and more holistic than the one usually taken, which tends to concentrate primarily on the program or task.

Giving due consideration to the broader context does not mean that CALL-specific software should immediately be discarded, or, if an institution has chosen, say, *BlackBoard* as its favored LMS, the language teacher should feel compelled to use it. However, it does mean that the language teacher should be aware of these choices made at the larger scale and consider the *possibility* of making use of these options. The advantage lies in the support networks then available and the likelihood of providing an improved life span for any CALL materials produced. The argument is centered on the importance of integration across these various axes or dimensions, when it is feasible and justified.

The choice of the word *integration* has been made advisedly, even though the use of this label has been questioned. As Chapelle (2000, p. 220) reported, Patrikis (1997) said that integration was a "false lead" in that once technology is genuinely incorporated into language teaching and learning, everything changes. This point was reinforced by Postman (1993, cited in Debski, 1997, p. 41), who maintained that technological change is "neither additive nor subtractive" but "ecological" in that "one significant change generates total change." Similarly, Papert (1993) noted that "computers serve best when they allow *everything* to change" (p. 146). Although we tend to agree with the underlying goal of technology innovation that is being expressed here, it must immediately be said that this is not easily accomplished in educational settings and institutions with well-developed cultures and practices.

This more holistic perspective has been approached from a number of angles. Levy, for instance, argued for the value of the systems concept in relation to CALL. In this view, the "elements of the system can *only* be conceptualised meaningfully if they are viewed as part of the whole" (Levy, 1997, p. 66). In other words, there is an interrelationship between every element in the system and the system as a whole. Consequently, any element that is not working properly within the system will affect the whole system and, furthermore, any external influences impacting on the system will influence each element within it, to a greater or lesser degree. The challenge with this perspective, of course, is to identify the key interacting elements and locate the effective boundaries of the system. Sometimes it is unclear where exactly one draws the line between what is and what is not exerting an influence, and what exactly constitutes the system in a particular educational or language-learning setting. Conceptual frameworks can be helpful here. Note that activity theory proposes the activity *system* as the basic unit of analysis, where the activity system comprises a dynamic network of interacting and interdependent elements with its own cultural history.

From another angle again, Bax (2000, 2003) considered the integration of CALL in terms of normalization and commented, "This concept is relevant to any kind of technological innovation and refers to the stage when the technology becomes invisible, embedded in everyday practice and hence 'normalised'" (2003, p. 23). Note that this idea is primarily of value to classroom teachers, not to the field of CALL as a whole (see chap. 10). Thinking along similar lines, Rodgers (2003) used the term *routinizing*, which "occurs when an innovation has become incorporated into the regular activities of the organization and has lost its separate identity" (p. 428). These terms echo the well-known idea of the transparency of technologies when fully incorporated into everyday life, both generally and in relation to CALL, when the user focus is on the task at hand rather than the technology being used to undertake the task (see also Levy, 1997; Pennington, 2004; Winograd & Flores, 1986).

In reflecting on these ideas—integration, ecological change, the idea of the system, and the goal of normalization—and what these terms might mean in formal, educational settings, integration might be said to be a precursor to ecological change. With integration, we are still made aware of the discrete technology or software elements that need to be brought together with more conventional elements to form an effective, working whole; also, the emphasis is still with the individual components and how they might best be combined.

This compartmental view—and the use of the term *integration*—has developed for a reason in educational settings that use new technology. Unlike the wider community, educational institutions have tended to approach technology innovation in a rather particular way.

TECHNOLOGY INNOVATION AND CHANGE: INSIDE AND OUTSIDE THE INSTITUTION

Technology innovation and change occur rather differently in educational institutions when compared to the community at large, where ecological change and normalization may be said to occur more readily. To give an example in the general community, in Australia or Japan (where the authors live), we can say that mobile phones have been generally adopted and accepted, and that ecological change and normalization have occurred as far as this technology and the new cultures surrounding it are concerned. This does not mean that everyone has a mobile phone, nor that everybody wants one. However, it does mean in general terms that the technology is affordable and accessible, that technology support is available when needed, and that it works (mostly). In general, for the more affluent countries in the Western world, normalization basically reduces to an acceptance, by a large number of people with sufficient disposable income to purchase the item, of a robust and practical technology that is widely perceived to be useful. The decision to adopt the technology is in most cases a personal one, and the user meets the costs, both in the initial purchase and in subsequent use.

In contrast, most educational institutions are obliged to approach technology innovation rather differently. Generally, institutions do not have the resources to cater directly for the specific needs and preferences of every individual student. They have to be more systematic and inclusive, and often an institutionwide policy and approach is adopted. As far as possible, the decisions they make must be perceived to be fair and equitable. Overall, they do not provide technology resources for some and not others in their institutions. Also, institutional planners have to be keenly aware of the costs involved and the technical support required, among many other factors. Issues such as security and expense often lead to computers being grouped collectively in a small number of fixed locations (the computer lab/room). Factors such as the number and availability of technical staff and the nature of warranty agreements often lead to limited and focused support for a certain limited set of hardware and software, with the regular phasing in (and out) of machines after set periods (e.g., a 3-year cycle). It is much easier—and certainly less costly—for the institution to manage a narrow consistency in hardware and software than it is to embrace individual choice and variation.

As a result, the institution's capacity to cater for specific individual characteristics, needs, and goals is necessarily limited. This applies to staff and students, as well as disciplines and content areas within the institution. Staff may be obliged to use a particular kind of computer and particular software applications. Similarly, students have to work within the confines of the boundaries that have been set. The decision making generally operates at a much higher level in the institution than the department or center, so that language teachers find themselves with certain technological resources at their disposal as a result of decision making in which they have played little or no part.

There are other important factors that have the same net effect. These typically concern the culture of educational institutions and their practices, which can be remarkably resistant to change. Although on the surface new technologies and software applications may appear to be embraced, in matters relating to their use often there may be very little evidence of significant innovation and change. New technology practices may make inroads into some aspects of courses and programs, but not others. Individual teachers will only be able to exert control over some of these factors.

A good example lies in the way educational institutions approach assessment and examinations as far as technology is concerned. More often than not, pen-and-paper examinations are still employed, as are very basic question types (e.g., multiple choice), especially in technology-oriented tests. Unfortunately for language learners, this situation often seems to arise with especially important internal and external examinations that really matter with regard to the students' future career prospects (e.g., international English proficiency examinations). Thus, although students may make imaginative use of new technologies in their coursework—creating Web sites, developing portfolios, completing Web quests, participating in collaborative projects—all too often they are still required, individually, to complete examinations and assessment items that use very traditional techniques and technologies.

There are other examples in which there are significant differences inside and outside the institution from the students' point of view (horizontal integration). Outside the institution, students may use sophisticated software (interactive, Web-based games), but employ very conventional software within it. Students may make regular use of mobile technologies in their lives outside class, but within the institution such technologies may be frowned on or prohibited altogether. Similarly, students may make virtually no use of pen and paper at home, but then be required to use them extensively in the institution, especially in examinations. This is not to say that general-purpose applications should necessarily be the same as pedagogical applications, nor is it to say that students should not develop expertise across boundaries (see Engeström, Engeström, & Kärkkäinen, 1995). However, opportunities for greater continuity need to be recognized and, where appropriate, exploited. When viewed through the students' eyes, needless differences in technology selection—in hardware, software, and the user interface—can be surprising (see Hémard, 2003), or, more importantly, they can be confusing and demotivating. There is also a price in terms of the time required for further skill acquisition and training, time that usually has to be carved from the time available in the regular language-learning class. We believe that educators need to aim at achieving a greater degree of continuity between the worlds of the student inside and outside the institution, and inside and outside the language class within the institution. This is why much was made of horizontal integration in chapter 2 of this book, which addressed design.

Many of these decisions and events militate against ecological change and normalization in educational settings. In many everyday educational settings, the *goal* may be ecological change and normalization, but the *real-*

ity is still more accurately described as incremental attempts at integration, especially as perceived by users, language teachers, and their students. The institutional leaders may argue that their systems are highly coherent and integrated, but often this view is based on a "one suit fits all" policy that does not recognize content area differences (i.e, the need for a language-learning-specific response to technology choice), let alone individual differences and preferences within a content area. At the same time, with present rates of funding, at least in Australian universities, we cannot be too hard on the institution. They themselves are forced to rationalize and make very difficult choices on what particular technologies to resource and support.

NORMALIZATION (IN THE INSTITUTION)

Bax (2003) argued that normalization should be the "end goal for CALL," and asserted that we must "plan for this normalised state and then move towards it—indeed this offers and structures our entire agenda for the future of CALL" (p. 24). Although we dispute the value of normalization for all CALL activity (see chap. 10), we certainly see its usefulness as a goal in guiding language-teaching practice with technology. Bax (2003) went on to suggest a three-step approach for achieving this goal:

> The first step is to identify the critical factors which normalisation requires. The second is to audit the practice of each teaching context in the light of these criteria; the final step is to adjust our current practice in each aspect so as to encourage normalisation. Following this procedure will give each institution and teacher a clear framework within which to audit progress, and within which any obstacles to integration and normalisation can be identified and dealt with. (p. 24)

> This will almost certainly require changes in technology, in the size, shape and position of the classroom computer. It will require change in attitudes, in approach and practice amongst teachers and learners; it will require fuller integration into administrative procedures and syllabuses. (p. 27)

The Critical Factors

The first step is to identify the critical factors that normalization requires. Building on the list by Bax (2003), a tentative start list for consideration might be:

1. Easy access to the appropriate technologies (hardware/software), when required.
2. Acceptance by administrators that language learning has particular hardware/software needs.
3. Reliable technologies and applications:

    a. Technical support when needed.
    b. CALL materials that are robust and easy to use.
4. Reliable and willing partners in collaborative projects.
5. Acceptance of CALL activity by staff and students *as normal practice*:
    a. CALL materials that are relevant to the goals and needs of the students.
    b. Training for staff and students.

Although we may be able to make informed guesses at what factors are likely to be more generally applicable, the relative influence or impact of individual factors will inevitably vary from place to place. Local issues will always play a very important part. Thus, the particular weighting of factors—the priority order of importance in any particular setting—is likely to vary and be highly context specific.

Unfortunately, many of the factors involved are likely to lie well beyond the control or the direct influence of the individual language teacher. Decisions concerning the location and distribution of computers within an institution that are highly likely to impact on normalization, for example, are not usually made by language teachers. Yet, questions of access are often a major concern. Also, institutions may only be able to provide general-purpose applications software and little support for other kinds of CALL materials; these include process materials, such as collaborative learning environments, that may require focused technical support or the lifting of certain firewall restrictions. In addition, levels of training and understanding of CALL vary widely. Without appropriate training, neither staff nor students can hope to incorporate CALL as normalized practice (see Hubbard, 2004b).

Aside from these more general observations, we also know that CALL is context specific. In any particular situation, certain factors will present themselves as pivotal concerns, whereas others will be of less immediate relevance or importance. In one setting, the question of access might be crucial; in another, a fixed and nonnegotiable curriculum might be a major barrier to innovation; in a third, teacher training and attitudes might be central. In each teaching context, the situation will be a little different and, as Bax (2003) suggested for Step 2, each situation will need to be considered on its merits before steps toward normalization can be taken.

In spite of such reservations, we believe that working toward normalization is a useful, practical strategy. Language teachers are very much working within a complex system of opportunity and constraint. Normalization then becomes a process of understanding the infrastructure, the support networks, and the materials, and working effectively within them. All these factors yet again point to the importance of design and evaluation, and the adaptation of means to preconceived ends (see chap. 10).

Types of Research and Goals

As far as the kind of research that is needed to support steps towards normalization, Bax (2003) stipulated that there is a need for more *in-depth ethnographic studies* to identify key factors and to examine and clarify the relationship between them. He also noted that there needs to be *action research* in specific learning environments to identify barriers to normalization and ways of overcoming them.

Conducting in-depth ethnographic studies and action research studies is generally appropriate given the goals, the scope, and the kind of data that are likely to be collected in these kinds of research studies, and the type of information that would appear to be necessary to move toward normalization. Perhaps the value of in-depth ethnographic studies might be questioned, given the time they take, the technology changes that will occur over this time period, and their lack of generalizability beyond the specific context. It may be that smaller-scale, focused, short-term evaluation studies will provide the information necessary to guide the teacher or administrator toward normalization. Sometimes, the barriers to normalization in specific teaching contexts are rather obvious (and sometimes insurmountable, at least for the individual classroom teacher), and thus do not require any research at all. In such cases, it may become a question of vision, policy, and political action more than anything else. Generally speaking, we would think that small-scale evaluation studies or action research studies would provide the most effective approach to gathering the information required to encourage and facilitate normalization.

In thinking about research in this context, systemic approaches are generally very useful. Salomon (1991) provided an excellent overview of their strengths in a paper on transcending the qualitative–quantitative debate. He advocated systemic approaches as being most effective in the study of "complex learning environments undergoing change," with the assumption that "elements are interdependent, inseparable, and even define each other in a transactional manner so that a change in one changes everything else and this requires the study of patterns, not single variables" (p. 10). This general orientation is very compatible with the ideas we have been discussing. This is further reinforced when Salomon (1991) commented that, with systemic approaches, the research is dealing with a "whole dynamic ecology" (p. 12), the "newly created classroom culture" (p. 13) and "authenticity" (p. 16). This fits very well with what we said earlier about normalization, and ecological change.

However, it would be an unfair representation of Salomon's paper to imply that systemic approaches were all that were needed. Salomon (1991) also asserted:

The systemic study of complex learning environments cannot be fruitful, and certainly cannot yield any generalisable (applicable) findings and conclusions, in the absence of carefully controlled analytic studies of selected aspects in which internal validity is maximised. (p. 16)

For one needs to know what aspects of the complex setting deserve to be studied in greater detail under controlled conditions. The sources of such knowledge are one's detailed and systematic observations of the complex phenomenon. Without observations of the whole system of interrelated events, hypotheses to be tested could easily pertain to the educationally least significant and pertinent aspects, a not too infrequent occurrence. (p. 17)

Of course, the notion of the system permeates systemic approaches to research. In any particular teaching context, the teacher first needs to identify the elements of the system, as far as CALL is concerned. The system then has to be examined from within, with consideration of how the various elements interact with one another, and then from outside, with consideration of what factors or elements are likely to impact on the system or disturb its equilibrium. Systems are dynamic and are subject to change, so formal and informal evaluative studies will need to be ongoing.

The Language Teacher-Designer

If normalization is to be accomplished, language teachers must play a central role. They have the demanding task of trying to understand the contextual opportunities and constraints and then, with their students, help build an effective learning environment. Clearly, language teachers are themselves designers, as noted in chapter 2. This point is often overlooked, especially in terms of the decision making that is required to create and sustain language learning. Much of what design is fundamentally about concerns the teacher-designers' grasp and understanding of the opportunities and constraints. All design contexts, even the most generously funded, have such parameters, and their qualities and characteristics need to be understood if designs are to succeed.

It is surprising that language teachers have not more frequently been considered as designers. This is a powerful concept and deserves more attention as a pivotal role. Teachers are reasonably well recognized as materials designers (see Tomlinson, 1998) and, to a certain extent, syllabus and curriculum designers, but these roles are rather restricted and are still treated in the literature on language teaching as being somewhat marginal. What does the teacher's role as designer mean in a more general sense?

Viewing the language teacher as a designer brings to the foreground some critical insights. The first and most important of these is that the language teacher, in creating a product or plan of action, operates within a set of interrelated constraints. Constraints, often associated with the limited time and resources available to the teacher and the student, typically include the number of contact hours predetermined for a course, lesson times and durations, preparation time, access to new technologies and to software, development budget, technical support, ancillary learning materials, and so on. All of these constraints, in one way or another, directly impact on the ways in which a design is initially conceptualized and then brought to fruition.

The key factor to good design is to be able to identify and to understand the impact of authentic constraints and to be able to work creatively within them. To conceptualize language teaching or CALL without constraints and to assume "ideal conditions"—as is so often the case with theoretically derived models of language teaching and learning—is often to miss the point as far as successful design is concerned in real educational settings. Reflecting on practice, but working without restrictions of any kind, really does not teach very much. That is why the best preservice and inservice professional development courses build authentic constraints into the tasks that trainee language teachers are required to complete (see Hubbard & Levy, 2006). Then, later, novice teachers will be in a much stronger position to operate within the constraints that will inevitably impinge on their work during their professional lives. It is only when the designer engages with the constraints that a deeper understanding of the true nature of the design problem becomes clear. If the idea of the teacher-designer is coupled with the idea of the learner-designer, there is even further opportunity for creative design in language teaching and learning.

CONCLUSION

Contemporary work in CALL has the capacity to inform language teaching more broadly (see also Coleman, 2005). Understanding the language teacher's role as a designer is important and helpful, not only in the role of materials designer but also in the broader role as designer of a whole environment for language learning. We know that all design requires working within boundaries, with opportunities and constraints. Understanding the opportunities and constraints in any given setting, and being able to allow this knowledge and understanding to inform practice, is an important contributing factor for a successful language teacher. When technology is involved as well, a further dimension has to be incorporated that involves the appropriate and timely use of technological resources. Therefore, recognizing the role of the language teacher as a designer brings important insights, not only for CALL, but also when reflecting on the practice of language teaching more generally.

Another area in which CALL can inform language teaching in a broader way is that of language-learning materials development. Examples from

CALL can greatly extend and enhance our understanding of language-learning materials. It is quite remarkable, in our view, that there was scarcely a mention of CALL materials in Tomlinson's (1998) book on materials development in language teaching. As we noted in the introduction to this book, following Breen et al. (1979), materials provide not only content but also guidelines or frameworks for learning; in other words, in Breen's terms, they can be *process* materials as well as *content* materials. We already know that CALL has enormous potential for providing content materials in the form of Web resources. However, CALL can equally provide process materials, not only in such forms as the word processor, but also in terms of the structure and sequencing of authentic tasks and activities that students can complete online as if they were native speakers of the language.

Emergent and Established CALL

There are undoubtedly more dimensions to CALL than those we have selected for our attention in this book. Even so, through viewing CALL from these different perspectives, we believe we have been able to achieve a clearer sense of the scope of CALL—which is both broader and at the same time more specialized and focused than is sometimes imagined—and reach a better understanding of some of its unique qualities and characteristics.

Among the dimensions, we would argue that design is the most important. CALL is fundamentally a design discipline, and this is what makes it distinctive. Design, as we understand the term, is not so much about the way things look as about the way things work. The word *design* may be defined in a number of ways, but a good general definition for our purposes here is to say that design is the "adaptation of means to a preconceived end" (*Macquarie Dictionary*, p. 253). Good design requires both an understanding of the means—the potentials, capabilities, and constraints—and the ends—the goal or purpose. In CALL, this requires an understanding of the technologies that may be used for language learning, especially in terms of matching capabilities of the tool to the purpose, as well as an ability to manage the learning environment to meet the desired goals, often with the active participation of students. It also presupposes that the designers work within a specific context that is defined by a certain set of opportunities and constraints.

CALL design is inseparable from evaluation. Whatever is constructed needs to be tested and evaluated in some way, informally if not more formally. In this book, design and evaluation are involved in many examples that have been described and illustrated, including language-learning tasks, exercises, Web sites, online courses, tutors and tools, and learning environments. As these CALL materials are created, they need to be checked and tested against principled criteria or in practice. This may involve for-mal or informal, formative or summative evaluation techniques. Thus, de-sign and evaluation are located at the core of CALL. In many ways, how best to design and evaluate CALL materials is the fundamental question for the field. The other dimensions included in this book—CMC, theory, research, practice, and technology—feed into this question by providing technology options to consider, and theories and models to potentially guide research and practice. Questions concerning design and evaluation largely define scope, interests, and approaches, and what goes into making CALL unique as a field of study. In this book, in the order and content of the chapters, we have made the case that design and evaluation are primary when thinking about CALL.

In surveying the field, it soon becomes clear that technology itself pro-vides the backdrop. Even though CALL is not primarily about technology, and its practitioners are not principally trained in this area (as a rule), the technologies available at any particular time permeate the thinking of those involved. We have seen throughout this book a wide range of examples of technologies used for language learning, from state-of-the-art examples such as online learning systems and mobile technologies to well-established mainstream technologies, such as e-mail or the word processor. This quality of technology innovation, development, and diffusion to the wider commu-nity—in terms of the path that new technologies take from the testing of the initial concept through to prototype systems and, in successful cases, the adoption of the technology in education and the wider world—is central to an understanding of CALL.

For the designer, language teacher, or researcher wishing to work in such a diverse field, clearly there are many possible points of entry. Much will depend on one's attitude toward and aptitude for technology. Some practitioners are very interested in the technology dimension, whereas oth-ers are much less so. Working with very new technologies can be frustrating, but it can also be rewarding when genuinely new ground is broken. Addi-tionally, at any given moment certain technologies are more robust and more widely accepted than others. Thus, some practitioners will choose to work with relatively new, less proven technologies, whereas others will choose to work with established, generic technologies that have been widely accepted. This distinction is an important one.

In order to reflect the difference between those who engage directly with newer technologies and those who do not, in this section and the one that follows a distinction is drawn between *emergent* and *established CALL*. These terms have been created not to indicate a hard and fast division—in fact, any imaginary line drawn between them is a moving one, as state-of-the-art technologies and techniques become more accepted and shift from emer-gent to established, if a technology is not discarded altogether. However, the distinction does point to the fact, often overlooked, that many in CALL are attempting to do things directly with the technology that have not been done before. There is a tendency in the literature, especially in the more recent literature, to think about technology and language learning only in terms of the use of established technologies, such as CMC technologies like e-mail and chat. But many others are trying to break new ground by engaging directly with the technology itself. We introduce the term *emergent CALL* as a descriptor for this work. In contrast, those working in established CALL use mainstream technologies that are employed in education and society more broadly. This includes the use of information and communication technologies (ICTs) and mediated communication technologies such as CMC. In established CALL, directly shaping the technology itself is not a principal concern.

EMERGENT CALL

Introduction

From its earliest times, often within very limited resources, some CALL practitioners have chosen to become directly involved with the technology itself. Although such work has sometimes received criticism, some of it well justified, those who criticize can at times appear to disengage themselves entirely from issues concerned with technology design, almost as if acceptance of the commercial, off-the-shelf product were the only possible alternative. It is easy to forget that commercial products, such as word processing or e-mail applications, are themselves designed—and they are designed for a particular purpose and audience, mostly *native speakers* of *English*. Murray (2000), cited in Pennington, (2004, p. 26), went so far as to say, "When speakers of languages other than English try to use their mother tongue online, they are hampered by a technology that was designed for English ...." Many in the CALL community are very much concerned with the design of technologies for languages other than English; this point in itself is a compelling one for the existence of CALL as a specialized field of study.

Christina Haas (1996), in her seminal work on the materiality of literacy, asserted that one of the goals of research studies involving technology should be "to look *at*, rather than *through*, technology; that is, technology itself—rather than merely technology's effects—should be an object of inquiry" (p. 69, italics added). She continued, "Any changes in literate behaviour that computers facilitate or mandate are neither inevitable nor invisible. Rather they are the result of actual features—in this case, visual and tactile—of the technology" (p. 72).

In emergent CALL, we are very interested in looking at the technology to see what it can do and what it cannot. When there are shortcomings, these may be addressed. This approach may involve revisiting well-established and accepted technologies and seeing if their features can be improved or refined for language-learning purposes. Such was the case in the electronic resources (ETR) project designed by Appel and Mullen (2000, 2002), described in chapter 2. This example shows how purposeful design can be achieved for second-language learners by creating an e-mail environment with features specifically introduced to suit language learners. Over a period of time and through working closely with student users, four versions of the ETR resources were developed. Such developments demonstrate an uncompromising attitude when the received commercial version of an application is viewed as unsatisfactory, especially from a pedagogical perspective. Although such examples are certainly ambitious, they are very much at the centre of the discipline of CALL, where the technological and the pedagogical design really do come together on an equal footing in the mind of the designer. Generic products like word processors and e-mail applications are designed with commercial needs in mind. Although vast resources poured into commercial software development generally ensure high-quality products, they do not ensure suitability for nonnative speakers and languages other than English.

In emergent CALL, practitioners directly engage with the technology itself; in other words, they are looking at the technology rather than through it (Haas, 1996). There has been much groundbreaking work in CALL in this regard, and it needs to be recognized. As far as emergent CALL is concerned, teacher–designers, developers, and researchers are currently looking closely at language-learning programs involving:

- Speech-recognition applications.
- Broadband audiovisual technologies.
- Online teaching systems (with human tutors).
- Intelligent tutors (ICALL—with computer tutors).
- Mobile technologies.
- Fine-grained design decisions (the optimal annotation).
- Hybrid solutions.
- New authoring tools and techniques.
- Compatibility of technologies (e.g., knowledge pooling, reusability issues).

The qualities of emergent CALL were captured rather well by Armitage and Bowerman (2002): "We are always looking for ways in which *emergent technology* may drive CALL in unexpected directions; and we are always on the lookout for ways in which technological limitations negatively affect pedagogical practice" (p. 29, italics added). The stance recommended in this quote requires the CALL designer to adopt the role of the critic. It is not simply a case of adopting the latest technologies automatically; a degree of scepticism is required. Thus, it is also a question of critical appraisal of an emergent technology, of evaluation, in the light of a particular context of use. Critique requires a careful assessment of limitations as well as opportunities, and this is especially important for technologies that are new and emerging. Such a role calls to mind the advice of Phillips (1985a), from the early days of CALL: "It is necessary to build up a cadre of professionals with the knowledge and skills to hold *a watching brief* on educational technology" (p. 118, italics added).

Key Areas and Examples

ASR—Annotations. With new applications, designers typically frame and contextualize the questions: Does it work? If so, why does it work? In other words, designers are interested first in making a general assessment of the viability and robustness of a new technology; and then in designing prototype CALL programs and pinpointing exactly what particular design elements are responsible for them working successfully. Essentially, then, evaluation studies lead to research studies. Thus, Eskenazi (1999) posed the question of whether the automated speech recognition technology has come far enough for systems to be able to teach pronunciation properly. Then, at a lower, more focused level of analysis, there are many examples of designers trying to decide which is the best alternative from a small number of options. Thus Jones (2003), in an experimental study, sought to determine the most effective kind of multimedia application from four alternatives (see chap. 6). The comparison is narrowly focused, and we have seen that comparative studies conceptualized in this way help designers to make decisions between alternatives within particular design contexts.

How Help Works—ICALL—User Response. If a number of design elements or resources are involved, designers want to know which are used and, if so, how often and in what way. For example, Pujolà (2002) looked at the relative effectiveness of different resources within a set of help facilities in the program called *ImPRESSions* (see chap. 6). In ICALL, the quality and relevance of the feedback is critical; increasingly, the quality of the feedback is judged in relation to language learners with particular attributes. Thus, authors like Toole and Heift (2002) have tracked, then categorized, learner behavior and, sought appropriate labels for it.

In this work, assessing student attitudes and perceptions is crucial, especially in relation to gauging real and perceived difficulties in accessing and using the program. Any innovation needs formative evaluation, and users need to be involved early in the process and at regular intervals later in the program's development. Data-gathering techniques may involve survey evaluation or research connected with specific programs and design features. Alternatively, they may involve tracking, as was evident in all three of the emergent CALL projects described in chapter 6 (Chun, 2001; Jones, 2003; Pujolà, 2002). Just as design and evaluation have permeated this book, so have the user/learner orientation and the crucial importance of obtaining user responses at all key stages in design.

Online Tuition. Online tuition for language learning sits well in the category of emergent CALL. It is still sufficiently new to be considered the exception rather than the rule in language teaching and learning. Also, in some ways, online tuition is still controversial ideologically, especially if conceived and designed as a stand-alone unit, not related to a mainstream course, and when some may have perceived this alternative as a replacement rather than an addition or enhancement. On the other hand, for distance learners and often for minority languages (because of lack of teachers or resources), online tuition may be the only alternative.

A good example, described in this book, is in the evolution and design of *Lyceum* (Hampel, 2003; Hampel & Hauck, 2004). There are many interesting aspects associated with the design and evaluation of this project, but one raised in the discussion is the way in which the creators took a pluralist approach with regard to theory. Such an approach again conveys the complexity of creating online learning environments when designers are taking pains to adhere to a very principled approach. This example also demonstrates rather well that emergent CALL is not only about emergent technologies (although, according to our definition, it should include them); it is also understood to include innovation on two fronts—one technological, one pedagogical. We are concerned with an interaction between the opportunities and limitations presented by new technologies and specific pedagogical goals within a new learning environment. Some have described this as the *fit* between the attributes of the technology and the goals of the pedagogical design (see Chapelle, 2003; Hubbard, 1996; Levy, 1997).

Programming—Authoring—Hybrids. Of course, emergent CALL includes those constructing and testing software that use programming and scripting languages such as *XML or JavaScript*, as well as those that use the more advanced capabilities of authoring tools or LMSs such as *BlackBoard*. It also includes *hybrid solutions* to design problems in which the limitations of one development tool are counterbalanced by the strengths of another in a design, so that the two in tandem provide a more effective learning environment. For example, we see the emergence of hybrid solutions to design problems such as that described by Strambi and Bouvet (2003), who combined *WebCT* with *Hot Potatoes* to create the particular combination of features that the authors required in their learning environment. Designers in this category are pushing the boundaries of what can be done with the current generation of authoring tools, and are inventing innovative ways of dealing with shortcomings when they are found. Also, there are indications in such examples that designers are acknowledging the importance of vertical integration, and the value of using institution-supported products—especially the institution-supported learning management system (LMS)—even when they have constraints (see Arneil & Holmes, 2003). Sometimes it is better to work with such products, if it is possible without too much compromise, because of the institutional infrastructure and support available. If there are satisfactory ways of circumventing problems using hybrid solutions, and significant deficiencies in the institution-supported product can be ameliorated/corrected, then the positives may be seen to outweigh the negatives. Such issues as weighing alternatives and accepting trade-offs are, of course, what design is all about.

General Qualities

From these examples we can note a number of characteristics of those working within emergent CALL. One can detect a dissatisfaction with the ways in which mainstream technologies and applications shape the learning environment, combined with a willingness to try to change it. One can also observe an "I wonder if …" mentality, as designers ponder new options and possibilities. As a result, designers often wish to make improvements to the "off-the-shelf" version and to make it more attuned to the needs of nonnative language speakers, whose particular characteristics and goals are often researched but inadequately represented in technology design overall. To achieve improvements, designers also make use of prior research findings from the fields of design and second-language learning, as well as carefully derived principles from best practice.

As a general rule, the evaluations and research designs associated with emergent CALL tend to be focused and narrowly conceived compared to established CALL. Generally, designers have specific questions they want to ask, or a small number of alternatives they want to compare, test and choose between. Emergent CALL tends more toward an analytic rather than a synthetic approach (see Salomon, 1991; see also the discussion later in this chapter). Emergent CALL also tends to employ a combination of quantitative and qualitative methods, and gathers data from a number of complementary sources in order to create the richest picture of the phenomena in focus. As a research endeavor, emergent CALL is best described as a limited set of narrower research and development agendas that aim to identify, isolate, and test the effectiveness of key design elements in a CALL environment. Typically, designers and researchers are interested in solving particular, recognized design problems, creating individualized feedback in tutorial CALL programs, or evaluating and testing new forms of learning environment, as we saw with *Lyceum*, for instance. Those who enter the world of emergent CALL are destined for a challenging and fascinating experience.

ESTABLISHED CALL

Introduction

Established CALL is more concerned with *what is* rather than *what might be* technologically. Practitioners in this group generally look *through*, rather than *at*, the technology, to evoke the words of Haas (1996). Established CALL involves technologies that are well established and accepted. The label is used to indicate mainstream activity in contrast to more specialized activity involving new and emerging technologies. Practitioners focus on using and evaluating CMC modes for language learning and, when CALL materials are developed, well-known authoring tools such as *Hot Potatoes* or *BlackBoard* are used in a straightforward way (i.e., without advanced adaptation).

To employ tried and tested technology is entirely understandable, especially given the array of new technologies and applications available, and the fact that at any particular time many remain untested for language learning. There is inevitably a time lag between the introduction of a new technology or application and the capacity of the language-teaching profession to absorb and evaluate it. Many in established CALL also believe that focusing directly on the technology is a distraction, drawing attention away from the main activity of language teaching. In our view, this illustrates an important difference in focus between the CALL professional and the language-teaching professional generally (see Hubbard, 2004a; Hubbard & Levy, 2006). The view taken here is that the concerns of practitioners working in this area of established CALL are simply different from those working in emergent CALL. We believe that both have a very important role to play.

It is also important to emphasize that established CALL practitioners are still breaking new ground: It is simply that they are not concerned directly with manipulating the technology itself. Instead, the innovation lies in the design and evaluation of language-learning tasks, projects, online courses, and exercises using established authoring tools, LMSs, or CMC modes. Those in established CALL are also very much concerned with practice and the successful integration of CALL activities into language learning. Additionally, they are deeply aware of evaluation issues because they want to make sure that the time expended using technology-related language-learning activities is effective and well spent.

The use of the term *integration* in the last paragraph provides a timely cue for a link to the previous chapter. The terms *integration* and *normalization* apply most especially to established CALL. They embody the idea of regular, everyday practice, with the technology component kept firmly in its place—in the background. Such concepts incorporate a dynamic process of adjustment to the prevailing features of the surrounding environment. This involves astute recognition of the factors that are exerting an influence, both positive and negative, and then managing these influences effectively. This process and the factors involved will vary from place to place. For example, whereas in one setting a greater level of normalization might be achieved by providing more computers, in another setting the priority might be learner training, a modified pedagogy, or a more integrated approach to the CALL and non-CALL aspects of the curriculum. In each situation, the language teacher has to make a judgment on the most outstanding issues or problems and work toward solving them one by one, as far as it is within the teacher's capacity to do so, until a greater equilibrium or balance is achieved. Although this goal may seem to remain just out of reach, and an aspiration rather than a reality, it is a worthy goal. Integration and normalization play an important role in established CALL.

Established CALL is represented in many publications concentrating on technology applications in language teaching, or the design of language-learning tasks; thus, it is this kind of CALL that features predominantly in the discussions of practice, pedagogy, and CMC in this book. This includes network-based language teaching (Warschauer & Kern, 2000), CMC-based CALL (Harrington & Levy, 2001), and web-enhanced language learning (Taylor & Gitsaki, 2003). According to Chapelle (2000), NBLT "represents an expansion rather than a reconceptualisation of CALL" (p. 222). In almost all cases in NBLT, the technologies involved are well established.

In established CALL, the concern is not with the technology itself. In fact, some may argue that this should be the standard, accepted position adopted across all CALL activity. We would counter by saying that this position is risky: It rarely challenges the adoption of off-the-shelf technologies designed primarily for native speakers of English. It does not recognize that the design of these technologies is not predetermined through some kind of immutable law, but rather is consciously targeted by selecting an option from a number of design alternatives to meet the specific needs and preferences of a selected audience. This is perhaps where the CALL specialist sees things a little differently from the language teacher, and another reason why the distinction between emergent and established CALL is useful. We believe that both roles and agendas are needed.

Evaluation plays an important, and rather different, role in established CALL. Here, the priority is to evaluate CALL activities in real, educational settings in which evaluation is fully contextualized and conducted, such as the language classroom. A range of checklists and evaluation frameworks, both general and more specific, have emerged to meet this need. In the approaches to evaluation, we have seen a steady development in sophistication and complexity leading to the Hubbard and Chapelle frameworks described in chapter 3.

We can say that those working broadly within the bounds of established CALL are concerned with the following:

- Task and activity design.
- Project design.
- Web site design.
- Evaluation.
- Integration.
- Classroom management.
- Teaching techniques.
- Introducing students to CALL.
- Materials and resources (online and offline).
- Assessment.
- CMC (research and practice).

Task design provides an important focal point for members of this group. The aim is for task and project construction in CALL to always proceed on a principled basis, using criteria drawn from theory, research, and best practice. Successful integration is also critical. For established CALL, many of the most pertinent issues can be discussed around the nature of language-learning tasks, and design and evaluation.

Language-Learning Tasks

In established CALL, language-learning task design is very much at the heart of the matter. Task design is a feature in many books and journal articles in CALL (e.g., Chapelle, 2001, 2003). Design and evaluation frameworks have been built around the task, much research has been undertaken with the task as the focal point, and task design and structure have been written about extensively in the literature. The centrality of the task in thinking about CALL was also emphasized in chapter 2 of this book, and projects and discussion relating to this topic have arisen repeatedly.

How exactly the task is defined and constituted, both in isolation and as part of a sequence or cycle, is a key question. Defining the language-learning task is by no means an easy matter, and there are alternative and conflicting views on how this should be done. There are also a number of associated and overlapping reasons why task design and evaluation is problematic for practicing language teachers using technology.

Design. From the 1980s until the present day, the concept of the task and, relatedly, task-based language teaching, have been defined and presented in a variety of ways. Beyond the general idea of tasks as meaning-based activities closely related to learners' actual communicative needs and with some real-world relationship, no single agreed-on position has been reached. Furthermore, there is no single task-based language teaching (TBLT) methodology, although most approaches using the label share a three-phase structure of pretask, task, and posttask, and the idea of the task cycle, which can incorporate different aims at different stages, is well recognized (see Levy & Kennedy, 2004; Willis, 1996). Also, like communicative language teaching (CLT), there are strong and weak versions of TBLT, depending on whether the task component is perceived as a necessary *and* sufficient, or a necessary *but not* sufficient "driver" for language development (see Klapper, 2003). The distinction hinges on whether tasks alone are enough for language acquisition to occur, or whether they must be supplemented with some form-focused instruction.

Klapper (2003) provided a useful critique of both TBLT and CLT before it. He himself argued for a weaker version of TBLT that is consistent with his belief that "tasks in the communicative classroom should be linked to a more consistent focus on form and more guided practice than is currently envisaged in 'strong' versions of task-based teaching" (p. 33). Klapper also identified a number of shortcomings of TBLT, especially in activities in which nonnative speakers are interacting with one another. For example, referring to Bruton (2002), he commented, "Another concern is that the type of negotiated oral interaction around which the task-based approach is built will inevitably result in learners being exposed to large amounts of non-native language input which will merely serve to confirm their current interlanguage representations (Bruton, 2002)" (p. 38). This concern is potentially important for CALL in CMC exchanges involving only non-native speakers.

Critiques such as those provided by Klapper (2003) are valuable, and we should remember that TBLT is not yet "bedded down" as an approach to language teaching. Doughty and Long (2003) described TBLT as an "embryonic theory of language teaching, not a theory of SLA" (p. 51). Clearly, TBLT is still evolving, so we should be cautious in making claims about the effectiveness of tasks as an all-encompassing strategy. In addition, we have seen that the concept of the language-learning task has evolved from one that focused primarily on developing the students' communicative ability to one that involves a much broader conceptualization. If we take Ribé and Vidal (1993) as a guide, we should be thinking about third-generation tasks with their goal of enriching the students' personal experience more widely. Viewed in this way, tasks are composite in nature and consist of a variety of elements, each of which may be said to correspond to particular learning goals. Thus, the modern language-learning task is not a simple, one-dimensional object. As a result, designing effective tasks is a challenging exercise, with or without technology.

Ribé and Vidal's (1993) description of the three generations of the language-learning task also reflects a developing understanding of the nature of the "language" to be learned. It is not self-evident what exactly constitutes the language that provides the objective in language learning. By way of example, Bachman (1990, p. 88) broke language down into language competencies. He identified six kinds of language competence (knowledge of language): organizational competence, comprising grammatical competence and textual competence; and pragmatic competence, comprising illocutionary and sociolinguistic competencies. He specified these various competencies in order to construct L2 learners' proficiency tests. For us, the point is that the language object can be interpreted in a variety of ways. In very simple terms, the language object and the goals of language learning may be conceived narrowly (e.g., developing grammatical competence), or more broadly (e.g., to include the range of competencies described by Bachman, and outward from there to include the development of learning strategies and learner autonomy in a wide variety of content areas, including the very specific.). This narrow or broad conception of language is echoed in narrow and broad conceptions of the language-learning tasks that are held to be the vehicles for language learning.

Such issues and concerns relating to the definition of language and the language-learning task carry over into CALL. We have seen this reflected in a number of discussions in this book, including Furstenberg et al. (2001), who emphasized the importance of culture and intercultural competence; Sengupta (2001), who stressed the personal and social dimensions of language learning; and Warschauer (2000a), who wrote of developing new literacy and life skills. Such views represent expanding and shifting priorities for language learning, and again reemphasize the points made earlier. We continue to see this widening of the language teacher's understanding of the language-learning task.

When we add to this mix the possibilities that CALL opens up in terms of task design, we see that there is a real danger of conceptualizing language-learning tasks in ways that are unnecessarily narrow. We saw in chapter 2 that the resources available via the Internet open up new ways to think about tasks, such as Web quests and Web inquiry projects, which enable the language to be manipulated in the service of broader goals. Similarly, the CMC modes enable different kinds of communication to occur in NNS–NNS and NNS–NS interactions. Again, in these environments, new kinds of language-learning tasks can be envisioned. It is important that tasks primarily conceived for the face-to-face context or for research purposes are not necessarily assumed to be appropriate in online environments, in which there are numerous opportunities to enlarge and expand on traditional ideas of the language-learning task.

Evaluation. In spite of these complications, we need evaluation frameworks and methodologies that may be used to evaluate CALL tasks. The Hubbard (1988, 1992, 1996) and Chapelle (2001) frameworks provide mechanisms for evaluation in CALL, especially Chapelle's framework, which is explicitly designed for the evaluation of tasks. This comprehensive framework is valuable, although it is important to remember the way that focus on form is prioritized through the definition of *language-learning potential*, and that other potential priorities for language learning are placed in a secondary position and packaged under *positive impact*.

In this framework, language learning is defined purely in terms of how to manipulate the forms of the language, which is considered to be the *core* task of learning a language. This definition reflects the position of the second-language acquisition researcher rather than the language teacher in the classroom, who is faced with many competing goals. It also does not capture the ways in which task goals may focus on a number of different language aspects over time (i.e., to account for many of the competencies that need to be acquired in the learning of a language), nor the opportunities available for reconceptualizing language-learning tasks in CALL environments. This definition is appropriate for research purposes, but we think it is less valid and appropriate for teaching purposes. It may prove to be unnecessarily restrictive.

For those in established CALL, the design and evaluation of language-learning tasks is challenging. At first glance, the complexities can appear overwhelming, but as practicing language teachers know, beginning with identifying the needs of the actual group of students being taught and understanding the local technology infrastructure and support systems assist greatly in providing an initial point of departure. Also, the checklists and frameworks for CALL evaluation provide not only useful lists of factors that might influence teaching and learning, they additionally contribute procedures that may be followed in order to make decisions, or to formulate more in-depth empirical research studies.

CONCLUSION

Emergent and established CALL *together* give a much more complete picture of current work in the field. Those working in emergent CALL are tackling broad issues concerning whole technologies, as well as more narrowly defined questions such as setting up experiments to choose between two or three possible design alternatives. They are, for example, assessing the effectiveness of automated speech recognition for nonnative speakers in a variety of CALL programs, and investigating the optimal way to combine technology modes in online, multimodal teaching and learning systems for distance learning. They are also examining how best to design help and feedback so that it is tuned to individual learner characteristics and behaviors, and considering which particular combination of annotations works most effectively. As in all CALL work, these efforts move well beyond the technology itself. Such design work involves weighing the pedagogical goals alongside the opportunities and constraints presented by the technology and the human resources available. It involves closely tracking students using programs and obtaining student feedback on key decisions. Throughout, design and evaluation questions are a central concern. Also, in their work, designers and researchers attempt to use a principled approach, sometimes drawing on appropriate theories, sometimes driven by best practice, in order to learn more about their task and to find effective solutions to CALL design problems.

In parallel, those working in established CALL are generally focused on questions of task design, CALL in the classroom, and evaluation and integration into a course or curriculum. Authoring tools and applications are favored for developing learning materials that are typically presented in the form of online, Web-based activities. Various forms of synchronous and asynchronous communication are also used with students in projects and activities of various kinds. Some involve NNS–NS interactions. In this work, identifying and designing appropriate tasks for learners is a recurring question, and one for which the opportunities for reconceptualizing the nature of online task-oriented learning are considerable. Again, design and evaluation are central. Those in established CALL are concerned with creating a coherent space for learning a language that students find conducive to the task. This involves keeping the technology aspect in the background—not always an easy task—so that students can primarily focus on language learning.

In sum, CALL, as a whole, in its emergent and established dimensions, has the capacity to inform, and be informed, by the disciplines and areas of research and practice that surround it. In this regard, we concur with the conclusions of Debski (2003) when he wisely noted that:

> Attempts to limit academic influences, either by insulating CALL (in terms of theoretical influence or even physical space), allying it too strongly with a single related discipline (for example, SLA), or by overemphasising its autonomy can all be seen as a disservice to the discipline. Research and academic programs in CALL may well thrive best when based in units with wide and encompassing agendas that are open to and capable of engagement in a

constructive dialogue with the variety of academic areas, across the arts to sciences spectrum. (This does not preclude CALL having its posts in other fields of inquiry.) (p. 184)

In many ways, perhaps the most interesting and challenging areas of academic endeavor in the contemporary world are both interdisciplinary and multidisciplinary. It is certainly important to keep the lines of communication open between cognate areas. Although CALL draws on many areas of knowledge and theoretical orientations outside its boundaries, it is also building a significant body of work that is uniquely its own, and it is developing research agendas to meet its needs and aspirations. This, we believe, is exactly as it should be as we continue to investigate the most effective ways to make use of new technologies in language learning.

Origins of This Book

This book began with a detailed analysis of a corpus of CALL work constructed from the published CALL literature in 1999. The results of an analysis of this body of work from the perspectives of research and design were reported in Levy (2000) and Levy (2002), respectively. Although the CALL corpus (1999) provided the initial impetus for this book, all subsequent issues of the four major journals mentioned in the next paragraph, and the majority of edited CALL volumes published before this manuscript was submitted in July 2005, have also been carefully reviewed for *CALL Dimensions*, in addition to books and journal articles in related areas.

Publications in 1999 were chosen initially to set this project in motion because of the particularly large number of CALL publications that year. The CALL corpus (1999) included, with a small number of exceptions, all the chapters in four books (Cameron, 1999a, 1999b; Debski & Levy, 1999; Egbert & Hanson-Smith, 1999) and all the articles in four major CALL journals: *Computer-Assisted Language Learning, CALICO, ReCALL,* and the online journal *Language Learning and Technology*. This amounted to 177 journal articles and book chapters in all, more than twice the number published in 2000. Not surprisingly, the articles varied widely in terms of goal, length, style, and audience, and included works focusing on research, theory, method, design, new technologies, evaluation, and practice.

Through the process of describing a large number of publications, CALL terminology was refined and, where appropriate, new terms were offered for debate. A unique and specially designed thesaurus of descriptors or set of indexing terms for this project was created and refined in a cyclical manner as the items in the CALL corpus (1999) were read, indexed, and abstracted. Considerable time was spent ensuring that their meanings were clear and unambiguous, and that also, collectively, they were sufficient to describe the articles in a detailed and appropriate way. The descriptors spanned all aspects of language teaching, learning, and technology pertaining to CALL research, development, evaluation, and practice. The descriptors were of two kinds: *candidate* descriptors (or *identifiers*) and straightforward descriptors. At the time this manuscript was completed, there were 177 descriptors in all (26 identifiers; 151 descriptors).

The identifier is a special kind of descriptor that is invoked in order to help provide a metalanguage for the description of CALL publications. The label *identifier* is borrowed from the thesaurus of ERIC descriptors (Houston, 1995, p. xx). Identifiers are candidate descriptors that deal with concepts not yet fully determined. Typically, identifiers are difficult to define or circumscribe precisely, at least initially. In the ERIC thesaurus, for instance, *freedom* and *happiness* are examples of former identifiers that now appear as fully fledged descriptors. The identifiers were intended to help classify the CALL article as a whole, and to capture the essential nature, content, and shape of the article. For this, the wording of the title was considered carefully, and particular attention was paid to the content of the introduction, conclusion, and any summary sections. Details concerning the research methodology of the project, the abstracting process, and the formulation of the identifiers and descriptors were given in Levy (2000, 2002).

The order of chapters of this book, with minor adjustments, has been set from the priorities revealed by the CALL corpus in the order of the most frequently applied CALL identifiers. The identifiers and their frequencies of occurrence are illustrated in Fig. A.1. Note also that to avoid unnecessary overlap, the *artifact* and *task* identifiers are incorporated into the remaining principal dimensions in the structure of this book rather than appearing as discrete items in themselves.

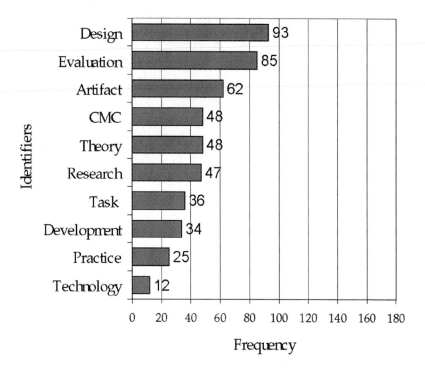

FIG. A.1.   The most frequent CALL identifiers (n = 177).

Features Matrix for the Main Studies Described in Chapter 6

1. Taylor and Gitsaki (2003)
   - General—WELL
   - Survey
   - Attitudes and perceptions
   - SLA: interaction account
   - Large scale
   - Cross-sectional, at intervals
   - 6-point, Likert scale

2. Tudini (2003)
   - Chat: NS–NNS
   - Distance learning
   - Without teacher
   - Chat logs
   - CMC research
   - Mediated versus face-to-face
   - Real-world versus classroom
   - Research informing pedagogy

3. O'Dowd (2003)
   - Intercultural language learning
   - E-mail: NNS–NNS (reciprocal)
   - Small scale
   - With teacher
   - Ethnographic
   - Action
   - Inductive
   - Longitudinal (1-year)
   - Task based
   - Network of methods

4. Chun (2001)
   - Reading, hypermedia
   - Qualitative/quantitative
   - Grounded theory
   - Individual differences
   - Tracking (look up behavior)
   - Strong materials/program descriptions
   - Network of methods
   - Correlational (in part)

5. Pujolà (2002)
   - Help design (2002)—assistance/ guidance
   - Feedback (Pujolà, 2001)
   - Tracking (ScreenCam)
   - Mixed theory base
   - Combination of methods

6. Jones (2003)
   - Listening, vocabulary
   - Annotations
   - A theory of multimedia learning (not language learning)
   - Experimental: 4 treatments
   - Statistical measures (MANOVA)

- Grounded theory—looking for patterns
- Detailed descriptive

- Mixed method
- Comparative
- Quantitative and qualitative
- Emergent data patterns: grounded (p. 46)
- Students' voice
- Explanation (p. 50)
- Individual differences (p. 55)
- Language: French (as opposed to English)

References

Abrams, Z. I. (2003a). Flaming in CMC: Prometheus' fire or Inferno's? *CALICO Journal, 20*(2), 245–260.

Abrams, Z. I. (2003b). The effect of synchronous and asynchronous CMC on oral performance in German. *The Modern Language Journal, 87*(2), 157–167.

Aitsiselmi, F. (1999). Second language acquisition through email interaction. *ReCALL, 11*(2), 4–11.

Aizawa, K., & Kiernan, P. (2003, September). *Cell phones in task-based learning: Are cell phones useful language learning tools?* Paper presented at the EUROCALL 2003 Conference, University of Limerick, Ireland.

Allum, P. (2002). CALL and the classroom: The case for comparative research. *ReCALL, 14*(1), 146–166.

Allwright, R. (1981). What do we want to teaching materials for? *English Language Teaching Journal, 36*(1), 5–18.

Al-Seghayer, K. (2001). The effect of multimedia annotation modes on L2 vocabulary acquisition: A comparative study. *Language Learning & Technology, 5*(1), 202–232.

Amichai-Hamburger, Y. (2005). Personality and the Internet. In Y. Amichai-Hamburger (Ed.), *The social net: Human behaviour in cyberspace* (pp. 27–56). Oxford, UK: Oxford University Press.

Anderson, J., Reder, L. M., & Simon, H. A. (1996). Situated learning and education. *Educational Researcher, 25*(4), 5–11.

Appel, C., & Gilabert, R. (2002). Motivation and task performance in a task-based web-based tandem project. *ReCALL, 14*(1), 16–31.

Appel, C., & Mullen, T. (2000). Pedagogical considerations for a web-based tandem language learning environment. *Computers and Education, 34*, 291–308.

Appel, C., & Mullen, T. (2002). A new tool for teachers and researchers involved in email tandem language learning. *ReCALL, 14*(2), 195–208.

Armitage, N., & Bowerman, C. (2002). Knowledge pooling in CALL: Programming and online language learning system for reusability, maintainability and extensibility. *Computer Assisted Language Learning, 15*(1), 27–54.

Arneil, S., & Holmes, M. (1999). Juggling hot potatoes: Decisions and compromises in creating authoring tools for the Web. *ReCALL, 11*(2), 12–19.

Arneil, S., & Holmes, M. (2003). Servers, clients, testing and teaching. In U. Felix (Ed.), *Language learning online: Towards best practice* (pp. 59–80). Lisse, The Netherlands: Swets & Zeitlinger.

Awaji, Y. (2004, June). *Constructing educational portal sites using free CMS: XOOPS and Moodle.* Paper presented at the JALTCALL 2004 Conference, Tokiwa University, Mito, Japan.

Ayres, R. (2002). Learner attitudes towards the use of CALL. *Computer Assisted Language Learning, 15*(3), 241–250.

Ayres, R. (2003). IELTS online writing—a team approach to developing an online course. *Computer Assisted Language Learning, 16*(4), 351–366.

Bachman, L. F. (1990). *Fundamental considerations in language testing.* Oxford, UK: Oxford University Press.

Bangs, P. (2003). Engaging the learner—how to author for best feedback. In U. Felix (Ed.), *Language learning online: Towards best practice* (pp. 81–96). Lisse, The Netherlands: Swets & Zeitlinger.

Barker, P., & King, T. (1993). Evaluating interactive multimedia courseware—a methodology. *Computers and Education, 21*(4), 307–319.

Barnes, S. B. (2003). *Computer-mediated communication: Human-to-human communication across the Internet.* Boston: Allyn & Bacon.

Barr, J. D., & Gillespie, J. H. (2003). Creating a computer-based language learning environment. *ReCALL, 15*(1), 68–78.

Barrette, C. M. (2001). Students' preparedness and training for CALL. *CALICO Journal, 19*(1), 5–36.

Barson, J. (1997). Space, time and form in the project-based foreign language classroom. In R. Debski, J. Gassin, & M. Smith (Eds.), *Language learning through social computing* (Occasional Papers Number 16; pp. 1–38). Melbourne, Australia: ALAA and the Horwood Language Centre.

Bateson, G. (2005, June). *Mashing hot potatoes with moodle: Tracking online quizzes with an open-source LMS.* Paper presented at the JALTCALL 2005 Conference, Ritsumeikan University, Japan.

Bax, S. (2000). Putting technology in its place. In K. Field (Ed.), *Issues in modern foreign languages teaching* (pp. 208–219). New York: Routledge Falmer.

Bax, S. (2003). CALL—past, present and future. *System, 31,* 13–28.

Beatty, K. (2003). *Teaching and researching computer-assisted language learning.* Hong Kong: Longman.

Beaudoin, M. (2004). Educational uses of databases in CALL. *Computer Assisted Language Learning, 17*(5), 497–516.

Beauvois, M. H. (1995). E-talk: Attitudes and motivation in computer-assisted classroom discussion. *Computers and the Humanities, 28,* 177–190.

Beauvois, M. H. (1997). Computer-mediated communication (CMC): Technology for improving speaking and writing. In M. D. Bush & R. M. Terry (Eds.), *Technology-enhanced language learning* (pp. 165–184). Lincolnwood, IL: National Textbook.

Beauvois, M. H. (1998). Conversations in slow motion: Computer-mediated communication in the foreign language classroom. *The Canadian Modern Language Review, 54*(2), 198–217.

Belz, J. A. (2001). Institutional and individual dimensions of transatlantic group work in network-based language teaching. *ReCALL, 13*(2), 213–231.

Benigno, V., & Trentin, G. (2000). The evaluation of online courses. *Journal of Computer Assisted Learning, 16,* 259–270.

Beretta, A. (1991). Theory construction in SLA: Complementarity and opposition. *Studies in Second Language Acquisition, 13*, 493–511.

Bernstein, J., Najmi, A., & Ehsani, F. (1999). Subarashi: Encounters in Japanese spoken language education. *CALICO Journal, 16*(3), 361–384.

Blackburn, S. (1994). *The Oxford dictionary of philosophy.* Oxford, UK: Oxford University Press.

Blake, R. J. (2000). Computer mediated communication: A window on L2 Spanish interlanguage. *Language Learning & Technology, 4*(1), 120–136.

Blin, F. (1999). CALL and the development of learner autonomy. In R. Debski & M. Levy (Eds.), *WORLDCALL: Global perspectives on computer-assisted language learning* (pp. 133–148). Lisse, Netherlands: Swets & Zeitlinger.

Blin, F. (2004). CALL and the development of learner autonomy: Towards an activity-theoretical perspective. *ReCALL, 16*(2), 377–395.

Block, D. (1996). Not so fast: Some thoughts on theory culling, relativism, excepted findings and the heart and soul of SLA. *Applied Linguistics, 17*(1), 63–83.

Blok, H., Van Daalen-Kapteijns, M. E., Otter, M. E., & Overmaat, M. (2001). Using computers to learn words in the elementary grades: An evaluation framework and a review of effect studies. *Computer Assisted Language Learning, 14*(2), 99–128.

Boyle, T. (1997). *Design for multimedia learning.* London: Prentice-Hall.

Brammerts, H. (1996). Tandem language learning via the Internet and the international e-mail tandem network. In D. Little & H. Brammerts (Eds.), *A guide to language learning in tandem via the Internet* (Occasional Paper Number 46; pp. 9–22). Dublin: CLCS.

Brandl, K. (2002). Integrating Internet-based reading materials into the foreign language curriculum: From teacher- to student-centered approaches. *Language Learning & Technology, 6*(3), 87–107.

Breen, M. P. (1986). Learner contributions to task design. In C. N. Candlin & D. F. Murphy (Eds.), *Language learning tasks* (pp. 23–46). Englewood Cliffs, NJ: Prentice-Hall.

Breen, M. P., Candlin, C. N., & Waters, A. (1979). Communicative materials design: Some basic principles. *RELC Journal, 10*(2), 1–13.

Brett, D. (2004). Computer generated feedback on vowel production by learners of English as a second language. *ReCALL, 16*(1), 103–113.

Brickmayer, J. D. (2000). Theory-based evaluation in practice. *Evaluation Review, 24*, 407–423.

Bruner, J. S. (1985). Vygotsky: A historical and conceptual perspective. In J. V. Wertsch (Ed.), *Culture, communication and cognition: Vygotskian perspectives* (pp. 1–32). Cambridge, UK: Cambridge University Press.

Brussino, G. (1999). Culture, communication, navigation and CALL: The role of the user interface and video material in a multimedia program for intermediate Italian learners. In K. Cameron (Ed.), *CALL & the learning community* (pp. 23–30). Exeter, UK: Elm Bank.

Brussino, G., Luciano, B., & Gunn, C. (1999). Integrated CALL design: Crescendo in Italia, a language teaching package for intermediate Italian learners. *Computer Assisted Language Learning, 12*(5), 409–426.

Bruton, A. (2002). When and how the language development in TBI? *ELT Journal, 56*(3), 296–297.

Burgess, J., & Spencer, S. (2000). Phonology and pronunciation in integrated language teaching and teacher education. *System, 28*, 191–215.

Burston, J. (2003a). Proving it works. *CALICO Journal, 20*(2), 219–226.

Burston, J. (2003b). Software selection: A primer on sources and evaluation. *CALICO Journal, 21*(1), 29–40.

Byram, M. (1997). *Teaching and assessing intercultural communicative competence.* Clevedon, UK: Multilingual Matters.

Cameron, K. (Ed.). (1999a). *CALL and the learning community.* Exeter, UK: Elm Bank.

Cameron, K. (Ed.). (1999b). *Computer assisted language learning (CALL): Media, design and applications.* Lisse, The Netherlands: Swets & Zeitlinger.

Cancelo, P. (2002). English Spanish interpreter deluxe. *CALICO Journal, 29*(2), 450–461.

Candlin, C. N. (1986). Towards task-based learning. In C. Candlin & D. Murphy (Eds.), *Language learning tasks.* (Lancaster Practical Papers in English Language Education, Vol. 7; pp. 5–22). London: Prentice-Hall.

Candlin, C. N., & Murphy, D. F. (1986). *Language learning tasks* (Lancaster Practical Papers in English Language Education, Vol. 7). London: Prentice-Hall.

Carey, M. (2004). CALL visual feedback for pronunciation of vowels: Kay Sona-Speech. *CALICO Journal, 21*(3), 571–601.

Carter, M. (1993). *Framing art: Introducing theory and the visual image.* Sydney: Hale and Iremonger.

Chambers, A. (2005). Integrating corpus consultation in language studies. *Language Learning & Technology, 9*(2), 111–125.

Chambers, A., & O'Sullivan, I. (2004). Corpus consultation and advanced learners' writing skills in French. *ReCALL, 16*(1), 158–172.

Chan, A., & Siu, M. (2005). Efficient computation of the frame-based extended union model and its application in speech recognition against partial temporal corruptions. *Computer Speech and Language, 19*(3), 301–319.

Chan, W. M., & Kim, D.-H. (2004). Towards greater individualization and process-oriented learning through electronic self-access: Project "e-daf." *Computer Assisted Language Learning, 17*(1), 83–108.

Chapelle, C. (1990). The discourse of computer-assisted language learning: Toward a context for descriptive research. *TESOL Quarterly, 24,* 199–225.

Chapelle, C. (1997). CALL in the year 2000: Still in search of research paradigms? *Language Learning & Technology, 1*(1), 19–43.

Chapelle, C. (1998). Multimedia CALL: Lessons to be learned from research on instructed SLA. *Language Learning & Technology, 2*(1), 22–34.

Chapelle, C. (1999a). Research questions for a CALL research agenda: A reply to Rafael Salaberry. *Language Learning & Technology, 3*(1), 108–113.

Chapelle, C. (1999b). Theory and research: Investigation of "authentic" language learning tasks. In J. Egbert & E. Hanson-Smith (Eds.), *CALL environments: Research, practice and critical issues* (pp. 101–115). Alexandria, VA: TESOL.

Chapelle, C. (2000). Is network-based learning CALL? In M. Warschauer & R. Kern (Eds.), *Network-based language teaching: Concepts and practice* (pp. 204–228). Cambridge, UK: Cambridge University Press.

Chapelle, C. (2001). *Computer applications in second language acquisition: Foundations for teaching, testing and research.* Cambridge, UK: Cambridge University Press.

Chapelle, C. (2003). *English language learning and technology: Lectures on teaching and research in the age of information and communication.* Amsterdam: John Benjamins.

Chiao, D. (1999). Using the Internet in English instruction at the Chinese Air Force Academy. In K. Cameron (Ed.), *CALL & the learning community* (pp. 39–50). Exeter, UK: Elm Bank.

Chikamatsu, N. (2003). The effects of computer usage on L2 Japanese writing. *Foreign Language Annals, 36,* 114–127.

Chun, D. M. (1998). Signal analysis software for teaching discourse intonation. *Language Learning & Technology, 2*(1), 61–77.

Chun, D. M. (2001). L2 reading on the web: Strategies for accessing information in hypermedia. *Computer-Assisted Language Learning, 14*(5), 367–404.

Chun, D. M., & Plass, J. L. (1997). Research on text comprehension in multimedia environments. *Language Learning & Technology, 1*(1), 60–81.

Chun, D. M., & Plass, J. L. (2000). Networked multimedia environments for second language acquisition. In M. Warschauer & R. Kern (Eds.), *Network-based language teaching: Concepts and practice* (pp. 151–170). Cambridge, UK: Cambridge University Press.

Cobb, T. (1999). Breadth and depth of lexical acquisition with hands-on concordancing. *Computer Assisted Language Learning, 12*(4), 345–360.

Cohen, E. (1994). *Designing groupwork: Strategies for the heterogeneous classroom.* New York: Teacher's College Press.

Coleman, J. A. (2005). CALL from the margins: Effective dissemination of CALL research and good practices. *ReCALL, 17*(1), 18–31.

Collot, M., & Belmore, N. (1996). Electronic language: A new variety of English. In S. Herring (Ed.), *Computer mediated communication: Linguistic, social and cross-cultural perspectives* (pp. 13–28). Amsterdam: John Benjamins.

Colpaert, J. (2004). *Design of online interactive language courseware: Conceptualisation, specification and prototyping.* Unpublished doctoral dissertation, University of Antwerp.

Coniam, D. (2004). Using language engineering programs to raise awareness of future CALL potential. *Computer Assisted Language Learning, 17*(2), 149–176.

Coniam, D., & Wong, R. (2004). Internet relay chat as a tool in the autonomous development of ESL learners' English language ability: An exploratory study. *System, 32*(3), 321–335.

Cook, V. (2001). *Second language learning and language teaching.* London: Hodder.

Cooper, A. (1999). *The inmates are running the asylum.* Indianapolis: Sams.

Corder, D., & Waller, G. (2005). An analysis of the effectiveness of a CALL software package for the learning and teaching of kanji and the development of autonomous language learning skills. *CALL-EJ Online, 7*(1). Retrieved December 1, 2005 from http://www.tell.is.ritsumei.ac.jp/callejonline/journal/7-1/

Cowan, R., Choi, H. E., & Kim, D.-H. (2003). Four questions for error diagnosis and correction in CALL. *CALICO Journal, 20*(3), 451–463.

Coyne, R. (1997). *Designing information technology in the information age: From method to metaphor.* Cambridge, MA: MIT Press.

Crompton, P. M. (1999). Integrating Internet-based CALL materials into mainstream language teaching. In K. Cameron (Ed.), *CALL and the learning community* (pp. 75–82). Exeter, UK: Elm Bank.

Crystal, D. (2001). *Language and the Internet.* Cambridge, UK: Cambridge University Press.

Cuban, L. (2001). *Oversold and underused: Computers in the classroom.* Cambridge, MA: Harvard University Press.

Curado Fuentes, A. (2003). The use of corpora and IT in a comparative evaluation approach to oral business English. *ReCALL, 15*(2),189–201.

Cushion, S. (2004). Increasing accessibility by pooling digital resources. *ReCALL, 16*(1), 41–50.

Cushion, S., & Hémard, D. (2003). Designing a CALL package for Arabic while learning the language ab initio. *Computer Assisted Language Learning, 16*(2–3), 259–266.

Cziko, G. A. (2004). Electronic tandem language learning (eTandem): A third approach to second language learning for the 21st century. *CALICO Journal, 22*(1), 25–40.

Cziko, G. A., & Park, S. (2003). Internet audio communication for second language learning: A comparative review of six programs. *Language Learning and Technology, 7*(1), 15–27.

Dabbagh, N., & Burton, L. (1999). The design, development, implementation and evaluation of a graduate level course for teaching web-based instruction. Retrieved December 1, 2005 from http://naweb.unb.ca/proceedings/1999/dabbagh/dabbagh.html

Dalby, J., & Kewley-Port, D. (1999). Explicit pronunciation training using automatic speech recognition. *CALICO Journal, 16*(3), 425–446.

Dalgarno, B. (2001). Interpretations of constructivism and consequences for computer assisted learning. *British Journal of Educational Technology, 32*(2), 193–194.

Daniels, P. (2004, June). *Tracking student progress online.* Paper presented at the JALTCALL 2004 Conference, Tokiwa University, Mito, Japan.

Danuswan, S., Nishina, K., Akahori, K., & Shimizu, Y. (2001). Development and evaluation of the Thai learning system on the web using natural language processing. *CALICO Journal, 19*(1), 67–88.

Darhower, M. (2002). Interactional features of synchronous CMC in the intermediate L2 class: A sociocultural case study. *CALICO Journal, 19*(2), 249–277.

Darhower, M. (2003). *CALICO Journal* software reviews: Connected speech. *CALICO Journal, 20*(3), 603–612.

Davies, G. (2001). New technologies and language learning: A suitable subject for research? In A. Chambers & G. Davies (Eds.), New technologies and language learning: A European perspective (pp. 13–27). Lisse, The Netherlands: Swets & Zeitlinger.

Davis, B., & Thiede, R. (2000). Writing into change: Style shifting in asynchronous electronic discourse. In M. Warschauer & R. Kern (Eds.), *Network-based language teaching: Concepts and practice* (pp. 87–120). Cambridge, UK: Cambridge University Press.

Debski, R. (1997). Support of creativity and collaboration in the language classroom: A new role for technology. In R. Debski, J. Gassin, & M. Smith (Eds.), *Language learning, through social computing* (Occasional Papers Number 16; pp. 41–65). Melbourne, Australia: ALAA and the Horwood Language Centre.

Debski, R. (2000). Exploring the re-creation of a CALL innovation. *Computer Assisted Language Learning, 13*(4–5), 307–332.

Debski, R. (2003). Analysis of research in CALL (1980–2000) with a reflection on CALL as an academic discipline. *ReCALL, 15*(2), 177–188.

Debski, R., & Gruba, P. (1999). A qualitative survey of tertiary instructor attitudes towards project-based CALL. *Computer Assisted Language Learning, 12*(3), 219–239.

Debski, R., & Levy, M. (Eds.). (1999). *WORLDCALL: Global perspectives on computer-assisted language learning.* Lisse, The Netherlands: Swets & Zeitlinger.

Decoo, W. (1984). An application of didactic criteria to courseware evaluation. *CALICO Journal, 2*(2), 42–46.

Decoo, W., & Colpaert, J. (2002). *Crisis on campus: Confronting academic misconduct.* Cambridge, UK: MIT Press.

De la Fuente, M. J. (2003). Is SLA interactionist theory relevant to CALL? A study on the effects of computer mediated interaction in L2 vocabulary acquisition. *Computer Assisted Language Learning, 16*(1), 47–81.

De Ridder, I. (1999). Are we still reading or just following links? How the highlighting or hyperlinks can influence incidental vocabulary learning. In K. Cameron (Ed.), *CALL & the learning community* (pp. 105–116). Exeter, UK: Elm Bank.

De Ridder, I. (2000). Are we conditioned to follow links? Highlights in CALL materials and their impact on the reading process. *Computer Assisted Language Learning, 13*(2), 183–195.

De Ridder, I. (2002). Visible or invisible links: Does the highlighting of hyperlinks affect incidental vocabulary learning, text comprehension, and the reading process? *Language, Learning, and Technology, 6*(1), 123–146.

Dias, J. (2002a). CELL phones in the classroom: Boon or bane? (Part 1). *Calling Japan, 10*(2), 16–22.

Dias, J. (2002b). CELL phones in the classroom: Boon or bane? (Part 2). *Calling Japan, 10*(3), 8–14.

DiMatteo, A. (1990). Under erasure: A theory for interactive writing in real time. *Computers and Composition, 7,* 71–84.

Dodge, B. (1995). Some thoughts about webquests. Retrieved December 1, 2005 from http://webquest.sdsu.edu/about_webquests.html

Dodigovic, M. (2002). Developing writing skills with a cyber-coach. *Computer Assisted Language Learning, 15*(1), 9–26.

Dodigovic, M. (2005). *Artificial intelligence in second language learning: Raising error awareness.* Clevedon: Multilingual Matters Ltd.

Donaldson, R. P., & Kötter, M. (1999). Language learning in cyberspace: Teleporting the classroom into the target culture. *CALICO Journal, 16*(4), 531–557.

Donato, R., & McCormick, D. (1994). A sociocultural perspective on language learning strategies: The role of mediation. *Modern Language Journal, 78*(4), 453–464.

Dörnyei, Z., & Skehan, P. (2003). Individual differences in second language learning. In C. J. Doughty & M. H. Long (Eds.), *The handbook of second language acquisition* (pp. 589–630). Oxford, UK: Blackwell.

Doughty, C. J. (1991). Theoretical motivations for IVD software research and development. In M. D. Bush, A. Slaton, M. Verano, & M. E. Slayden (Eds.), *Interactive videodisc: The "why" and the "how"* (CALICO Monograph Series, 2; pp. 1–14). Provo, UT: Brigham Young Press.

Doughty, C. J., & Long, M. H. (2003). Optimal psycholinguistic environments for distance foreign language learning. *Language Learning & Technology, 7*(3), 50–80.

Doughty, C. J. & Williams, J. (1998). *Focus on form in classroom second language acquisition.* Cambridge, UK: Cambridge University Press.

Dreyer, C., & Nel, C. (2003). Teaching reading strategies and reading comprehension within a technology enhanced learning environment. *System, 31,* 349–365.

Dron, J. (2003). The Blog and the Borg: A collective approach to e-learning. World *Conference on E-Learning in Corp., Govt., Health, & Higher Ed, 1,* 440–443.

Dudeney, G. (2000). *The Internet and the language classroom: A practical guide for teachers.* Cambridge, UK: Cambridge University Press.

Dusquette, L., & Barrière, C. (2001). Reading comprehension: CALL and NLP. In K. Cameron (Ed.), *CALL—the challenge of change: Research and practice* (pp. 97–106). Exeter, UK: Elm Bank.

Egbert, J. (2003). A study of flow theory in the foreign language classroom. *The Modern Language Journal, 87*(4), 499–518.

Egbert, J. (2005). *CALL essentials: Principles and practice in CALL classrooms.* Arlington, VA: TESOL.

Egbert, J., Chao, C., & Hanson-Smith, E. (1999). Computer-enhanced language learning environments: An overview. In J. Egbert & E. Hanson-Smith (Eds.), *CALL environments: Research, practice and critical issues* (pp. 1–13). Alexandria, VA: TESOL.

Egbert, J., & Hanson-Smith, E. (Eds.). (1999). *CALL environments: Research, practice, and critical issues.* Alexandria, VA: TESOL.

Egbert, J., & Thomas, M. (2001). The new frontier: A case study in applying instructional design for distance teacher education. *Journal of Technology and Teacher Education, 9*(3), 391–405.

Ehrman, M. E., Leaver, B. L., & Oxford, R. L. (2003). A brief overview of individual differences in second language learning. *System, 31*(3), 313–330.

Ellis, R. (1994). *The study of second language acquisition.* Oxford, UK: Oxford University Press.

Ellis, R. (1997). *SLA research and language teaching.* Oxford, UK: Oxford University Press.

Ellis, R. (2003). *Task-based language learning and teaching.* Oxford, UK: Oxford University Press.

Emde, S. V. D., Schneider, J., & Kötter, M. (2001). Technically speaking: Transforming language learning through virtual learning environments (MOOs). *The Modern Language Journal, 85*(2), 210–225.

Engeström, Y. (1987). *Learning by understanding: An activity-theoretical approach to developmental research.* Helsinki: Orienta-Konsultit.

Engeström, Y. (1999). Activity theory and individual and social transformation. In Y. Engeström, R. Miettinen, & R.-L. Punamäki (Eds.), *Perspectives on activity theory* (pp. 19–38). Cambridge, UK: Cambridge University Press.

Engeström, Y., Engeström, R. & Kärkkäinen, M. (1995). Polycontextuality and boundary crossing in expert cognition: Learning and problem solving in complex work activities. *Learning and Instruction, 5,* 311–336.

Ercetin, G. (2003). Exploring ESL learners' use of hypermedia reading glosses. *CALICO Journal, 20*(2), 261–284.

Erdogan, H., Sarikaya, R., Chen, S. F., Gao, Y., & Picheny, M. (2005). Using semantic analysis to improve speech recognition performance. *Computer Speech and Language, 19*(3), 321–343.

Eskenazi, M. (1999). Using a computer in foreign language pronunciation training. *CALICO Journal, 16*(3), 447–470.

Fallows, J. (2002). He's got mail. *New York Review of Books, 49*(4). Retrieved February 10, 2004 from http://www.nybooks.com/articles/15180

Farghaly, A. (Ed.). (2003). *Handbook for language engineers.* Stanford, CA: CSLI.

Farrington, B. (1986). 'Triangular mode' working: The Littre project in the field. *System, 14*(2), 199–204.

Felix, U. (2001). The web's potential for language learning: The student's perspective. *ReCALL, 13*(1), 47–58.

Felix, U. (2002). The web as a vehicle for constructivist approaches in language teaching. *ReCALL, 14*(1), 2–15.

Felix, U. (Ed.). (2003a). *Language learning online: Towards best practice.* Lisse, The Netherlands: Swets & Zeitlinger.

Felix, U. (2003b). Pedagogy on the line: Identifying and closing the missing links. In U. Felix (Ed.), *Language learning online: Towards best practice* (pp. 148–170). Lisse, The Netherlands: Swets & Zeitlinger.

Fernández-Garcia, M., & Martínez-Arbelaiz, A. (2002). Negotiation of meaning in non-native speaker synchronous discussions. *CALICO Journal, 19*(2), 279–294.

Fischer, G. (1998). *Email in foreign language teaching: Toward the creation of virtual classrooms.* Tuebinger: Stauffenburg Medien.

Fischer, R. (1999). Computer applications and research agendas: Another dimension in professional advancement. *CALICO Journal, 16*(4), 559–571.

Fleming, S., & Hiple, D. (2004). Distance education to distributed learning: Multiple formats and technologies in language instruction. *CALICO Journal, 22*(1), 63–82.

Fotos, S. (2004). Writing as talking: E-mail exchange for promoting proficiency and motivation in the foreign language classroom. In S. Fotos & C. M. Browne (Eds.), *New perspectives on CALL for second language classrooms* (pp. 109–130). Mahwah, NJ: Lawrence Erlbaum Associates.

Furstenberg, G., Levet, S., English, K., & Maillet, K. (2001). Giving a virtual voice to the silent language of culture: The *Cultura* project. *Language Learning & Technology, 5*(1), 55–102.

Ganderton, R. (1999). Interactivity in L2 web-based reading. In R. Debski & M. Levy (Eds.), *WORLDCALL: Global perspectives on computer-assisted language learning* (pp. 49–66). Lisse, The Netherlands: Swets & Zeitlinger.

Garrett, N. (1991). CARLA comes to CALL. *Computer Assisted Language Learning, 4*(1), 41–45.

Gass, S. M. (2003). Input and interaction. In C. J. Doughty & M. H. Long (Eds.), *The handbook of second language acquisition* (pp. 224–255). Oxford, UK: Basil Blackwell.

Gass, S. M., & Varonis, E. M. (1994). Input, interaction and second language production. *Studies in Second Language Acquisition, 16,* 283–302.

Gettys, S., Imhof, L. A., & Kautz, J. O. (2001). Computer-assisted reading: The effect of glossing format on comprehension and vocabulary retention. *Foreign Language Annals, 34*(2), 91–106.

Gherardi, S., Nicolini, D., & Odella, F. (1998). Toward a social understanding of how people learn in organizations: The notion of situated curriculum. *Management Learning, 29*(3), 273–298.

Gibbs, W., Graves, P. R., & Bernas, R. S. (2001). Evaluation guidelines for multimedia courseware. *Journal of Research on Technology in Education 34*(1), 2–17.

Gibson, J. J. (1979). *The ecological approach to perception.* London: Houghton Mifflin.

Gillespie, J. H., & Barr, D. (2002). Resistance, reluctance and radicalism: A study of staff reaction to the adoption of CALL & IT in modern language departments. *ReCALL, 14*(1), 120–132.

Gimeno-Sanz, A. (2002). E-language learning for the airline industry. *ReCALL, 14*(1), 47–57.

Glendinning, E., & Howard, R. (2003). Lotus ScreenCam as an aid to investigating student writing. *Computer Assisted Language Learning, 16*(1), 31–46.

Godwin-Jones, R. (2003). Emerging technologies: Blogs and wikis: Environments for on-line collaboration. *Language Learning & Technology, 7*(2), 12–16.

Godwin-Jones, R. (2004). Emerging technologies: Learning objects: Scorn or SCORM? *Language Learning & Technology, 8*(2), 7–12.

González-Bueno, M., & Pérez, L. C. (2000). Electronic mail in foreign language writing: A study of grammatical accuracy, and quality of language. *Foreign Language Annals, 33*(2), 189–197.

González-Lloret, M. (2003). Designing task based CALL to promote interaction: En busca de Esmeraldas. *Language Learning & Technology, 7*(1), 86–104.

Goodfellow, R. (1995). A review of the types of CALL programs for vocabulary instruction. *Computer Assisted Language Learning, 8*(2–3), 205–226.

Goodfellow, R. (1999). Evaluating performance, approach and outcome. In K. Cameron (Ed.), *CALL: Media, design and applications* (pp. 109–140). Exeter, UK: Elm Bank.

Goodfellow, R., Manning, P., & Lamy, M.-N. (1999). Building an online open and distance language learning environment. In R. Debski & M. Levy (Eds.), *WORLDCALL: Global perspectives on computer-assisted language learning* (pp. 267–286). Lisse, The Netherlands: Swets & Zeitlinger.

Gray, R., & Stockwell, G. R. (1998). Using computer-mediated communication for language and culture acquisition. *On-CALL, 12*(3), 2–9.

Greaves, C., & Yang, H. (1999). A vocabulary-based language learning strategy for the Internet. In R. Debski & M. Levy (Eds.), *WORLDCALL: Global perspectives on computer-assisted language learning* (pp. 67–84). Lisse, The Netherlands: Swets & Zeitlinger.

Green, A., & Youngs, B. E. (2001). Using the Web in elementary French and German courses: Quantitative and qualitative study results. *CALICO Journal, 19*(1), 89–124.

Gregg, K. R. (1993). Taking explanation seriously; or, let a couple of flowers bloom. *Applied Linguistics, 14*(3), 276–94.

Gregg, K. R. (2000). Review article. A theory for every occasion: Postmodernism and SLA. *Second Language Research, 16*(4), 383–399.

Gregg, K. R. (2003). SLA theory: Construction and assessment. In C. J. Doughty & M. H. Long (Eds.), *The handbook of second language acquisition* (pp. 831–865). Oxford, UK: Blackwell.

Groot, P. (2000). Computer assisted second language vocabulary acquisition. *Language Learning & Technology, 4*(1), 60–81.

Guillot, M.-N. (2002). Corpus based work and discourse analysis in FL pedagogy: A reassessment. *System, 30*, 15–32.

Gutiérrez, G. A. G. (2003). Beyond interaction: The study of collaborative activity in computer-mediated tasks. *ReCALL Journal, 15*(1), 94–112.

Haas, C. (1996). *Writing technology: Studies on the materiality of literacy.* Mahwah, NJ: Lawrence Erlbaum Associates.

Hall, K. (1996). Cyberfeminism. In S. Herring (Ed.), *Computer-mediated communication: Linguistic, social and cross-cultural perspectives* (pp. 147–172). Amsterdam: John Benjamins.

Halliday, M. A. K. (1993). Towards a language-based theory of learning. *Linguistics and Education, 5*(2), 93–116.

Hampel, R. (2003). Theoretical perspectives and new practices in audio-graphic conferencing for language learning. *ReCALL, 15*(1), 21–36.

Hampel, R., & Baber, E. (2003). Using Internet-based audio-graphic and video conferencing for language teaching and learning. In U. Felix (Ed.), *Language learning online: Towards best practice* (pp. 171–191). Lisse, The Netherlands: Swets & Zeitlinger.

Hampel, R., & Hauck, M. (2004). Towards an effective use of audioconferencing in distance language courses. *Language Learning & Technology, 8*(1), 66–82

Hardison, D. M. (2004). Generalisation of computer-assisted prosody training: Quantitative and qualitative findings. *Language Learning & Technology, 8*(1), 34–52.

Hardisty, D., & Windeatt, S. (1989). *Computer-assisted language learning.* Oxford, UK: Oxford University Press.

Harless, W. C., Zier, M. A., & Duncan, R. C. (1999). Virtual dialogues with native speakers: The evaluation of an interactive multimedia method. *CALICO Journal, 16*(3), 313–338.

Harrington, M. W., & Levy, M. (2001). CALL begins with a "C": Interaction in computer-mediated language learning. *System, 29*(1), 15–26.

Hawisher, G. E., & Selfe, C. L. (1998). Reflections on computers and composition studies at the century's end. In A. Snyder (Ed.), *Page to screen: Taking literacy into the electronic era* (pp. 3–19). New York: Routledge.

Haworth, W., & Cowling, D. (1999). The WELL project: Local participation and national evaluation. In K. Cameron (Ed.), *CALL and the learning community* (pp. 161–167). Exeter, UK: Elm Bank Publications.

Healey, D. (1999). Classroom practice: Communicative skill-building tasks in CALL environments. In J. Egbert & E. Hanson-Smith (Eds.), *CALL environments: Research, practice and critical issues* (pp. 116–136). Alexandria, VA: TESOL.

Healey, D., & Klinghammer, S. J. (2002). Constructing meaning with computers [Special Issue]. *TESOL Journal, 11*(3), 3.

Heift, T. (2001). Errors specific and individualised feedback in a web-based language tutoring system: Do they read it? *ReCALL, 13*(1), 99–109.

Heift, T. (2002). Learner control and error correction in ICALL: Browsers, peekers, and adamants. *CALICO Journal, 19*(2), 295–313.

Heift, T. (2003). Multiple learner errors and meaningful feedback: A challenge for ICALL systems. *CALICO Journal, 20*(3), 533–548.

Heift, T., & Nicholson, D. (2001). Web delivery of adaptive and interactive language tutoring. *International Journal of Artificial Intelligence in Education, 14*, 310–325.

Heift, T., & Schulze, M. (2003). Error diagnosis and error correction in CALL. *CALICO Journal, 20*(3), 433–436.

Hémard, D. (1997). Design principles and guidelines for authoring hypermedia language learning applications. *System, 25*(1), 9–27.

Hémard, D. (1999). A methodology for designing student-centred hypermedia CALL. In R. Debski & M. Levy (Eds.), *WORLDCALL: Global perspectives on computer-assisted language learning* (pp. 215–228). Lisse, The Netherlands: Swets & Zeitlinger.

Hémard, D. (2003). Language learning online: Designing towards user acceptability. In U. Felix (Ed.), *Language learning online: Towards best practice* (pp. 21–43). Lisse, The Netherlands: Swets & Zeitlinger.

Hémard, D., & Cushion, S. (2001). Evaluation of a web-based language learning environment: The importance of a user-centred design approach for CALL. *ReCALL, 13*(1), 15–31.

Hémard, D., & Cushion, S. (2002). Sound authoring on the web: Meeting the user's needs. *Computer Assisted Language Learning, 15*(3), 281–294.

Herring, S. (Ed.). (1996). *Computer-mediated communication: Linguistic, social and cross-cultural perspectives.* Amsterdam: John Benjamins.

Hew, S.-H., & Ohki, M. (2001). A study on the effectiveness and usefulness of animated graphical annotation in Japanese CALL. *ReCALL, 13*(2), 245–260.

Hew, S.-H., & Ohki, M. (2004). Effect of animated graphic annotations and immediate visual feedback in aiding Japanese pronunciation learning: A comparative study. *CALICO Journal, 21*(2), 397–420.

Hewer, S., Kötter, M., Rodine, C., & Shield, L. (1999, September). *The right tools for the job: Criteria for the choice of tools in the design of a virtual, interactive environment for distance language learners and their tutors.* Paper presented at CAL '99: Virtuality in Education, London.

Hickman, L. A. (1992). *John Dewey's pragmatic technology.* Bloomington: Indiana University Press.

Hillier, V. (1990). Integrating a computer lab into an ESL program. *CAELL Journal, 1*(2), 23–24.

Hincks, R. (2003). Speech technologies for pronunciation, feedback and evaluation. *ReCALL, 15*(1), 3–20.

Holec, H. (1981). *Autonomy in foreign language learning*. Oxford, UK: Pergamon.

Holland, V. M. (1999). Tutors that Listen. *CALICO Journal, 16*(3), 245–250.

Holland, V. M., Kaplan, J. D., & Sabol, M. A. (1999). Preliminary tests of language learning in a speech-interactive graphics microworld. *CALICO Journal 16*(3), 339–360.

Holland, V. M., Kaplan, J. D., & Sams, M. R. (Eds.). (1995). *Intelligent language tutors: Theory shaping technology*. Mahwah, NJ: Lawrence Erlbaum Associates.

Hope, G. R., Taylor, H. F., & Pusack, J. P. (1984). *Using computers in teaching foreign languages*. New York: Harcourt Brace Jovanovich.

Hopkins, J. (2004). Plagiarism in the virtual language classroom. *TEL & CAL, 3*, 6–15.

Hoshi, M. (2003). Examining a mailing list in an elementary Japanese language class. *ReCALL, 15*(2), 217–236.

Hosoya, Y. (2004, June). *Jisedai-no e-learning shisutemu "WebOCM"-o riyou shita CALL jugyou no jissai [The realization of CALL classes using a next generation e-learning system "WebOCM"]*. Paper presented at the JALTCALL2004 Conference, Tokiwa University, Mito, Japan.

Houser, C., & Thornton, P. (2004, June). *Writing with mobile devices: Success or failure?* Paper presented at the JALTCALL 2004 Conference, Tokiwa University, Mito, Japan.

Houser, C., Thornton, P., Yokoi, S., & Yasuda, T. (2001). Learning on the move: Vocabulary study via mobile phone email. *ICCE 2001 Proceedings*, 1560–1565.

Houston, J. E. (1995). *Thesaurus of ERIC descriptors*. Phoenix, AR: Oryx.

Hoven, D. (1999). A model for listening and viewing comprehension in multimedia environments. *Language Learning & Technology, 3*(1), 88–103.

Hoven, D. (2004, July). *Hands-on at a distance: Technology and alternative delivery*. Paper presented at the AMEP National Conference 2004, Charles Darwin University, Australia.

Hubbard, P. (1987). Language teaching approaches, the evaluation of CALL software, and design indications. In W. Flint Smith (Ed.), *Modern media in foreign language education: Theory and implementation* (pp. 227–254). Lincolnwood, IL: National Textbook.

Hubbard, P. (1988). An integrated framework for CALL courseware evaluation. *CALICO Journal, 6*(2), 51–72.

Hubbard, P. (1992). A methodological framework for CALL courseware development. In M. C. Pennington & V. Stevens (Eds.), *Computers in applied linguistics: An international perspective* (pp. 39–65). Clevedon, UK: Multilingual Matters.

Hubbard, P. (1996). Elements of CALL methodology: Development, evaluation and implementation. In M. Pennington (Ed.), *The Power of CALL* (pp. 15–32). Houston: Athelstan.

Hubbard, P. (2004a). Guest editorial. *Computer Assisted Language Learning, 17*(1), 1–6.

Hubbard, P. (2004b). Learner training for effective use of CALL. In S. Fotos & C. Browne (Eds.), *New perspectives on CALL for second language classrooms* (pp. 45–68). Mahwah, NJ: Lawrence Erlbaum Associates.

Hubbard, P., & Bradin Siskin, C. (2004). Another look at tutorial CALL. *ReCALL, 16*(2), 448–461.

Hubbard, P., & Levy, M. (Eds.). (2006). *Teacher education in CALL*. Amsterdam: John Benjamins.

Hudson, J. M., & Bruckman, A. S. (2002). IRC Français: The creation of an Internet based SLA community. *Computer Assisted Language Learning, 15*(2), 109–134.

Hughes, J., McAvinia, C., & King, T. (2004). What really makes students like a web site? What are the implications for designing web-based language learning sites? *ReCALL, 16*(1), 85–102.

Hunter, D., Watt, A., Rafter, J., Duckett, J., Ayers, D., Chase, N., Fawcett, J., Gaven, T., & Patterson, B. (2004). *Beginning XML* (3rd Ed.). Indianapolis, IN: Wiley.

Hutchby, I. (2001). *Conversation and technology: From the telephone to the Internet*. Oxford, UK: Basil Blackwell.

Hwu, F. (2003). Learner's behaviours in computer-based input activities elicited through tracking technologies. *Computer Assisted Language Learning, 16*(1), 5–30.

Iskold, L. V. (2003). Building on success, learning from mistakes: Implications for the future. *Computer Assisted Language Learning, 16*(4), 295–328.

Itakura, H. (2004). Changing cultural stereotypes through email assisted foreign language learning. *System, 32*, 37–51.

Jacobson, M. (1994). Issues in hypertext and hypermedia research: Toward a framework for linking theory-to-design. *Journal of Educational Multimedia and Hypermedia, 3*(2), 141–154.

Jager, S. (2001). From gap-filling to filling the gap. In A. Chambers & G. Davies (Eds.), *ICT and language learning: A European perspective* (pp. 101–110). Lisse, The Netherlands: Swets & Zeitlinger.

Jamieson, J., Chapelle, C., & Preiss, S. (2004). Putting principles into practice. *ReCALL, 16*(2), 396–415.

Jamieson, J., Chapelle, C., & Preiss, S. (2005). CALL evaluation by developers, a teacher, and students. *CALICO Journal, 23*(1), 93–138.

Jarvis, H. (2001). Internet usage of English for Academic Purposes courses. *ReCALL, 13*(2), 206–212.

Johanyak, M. F. (1997). Analyzing the amalgamated electronic text: Bringing cognitive, social, and contextual factors of individual language users into CMC research. *Computers and Composition, 14*(1), 91–110.

Johnson, D. M. (1992). *Approaches to research in second language learning*. New York: Longman.

Johnson, E. M. (2002). The role of computer-supported discussion for language teacher education: What do the students say? *CALICO Journal, 20*(1), 59–79.

Johnson, K. (2003). *Designing language teaching tasks*. New York: MacMillan.

Jones, A., & Mercer, N. (1993). Theories of learning and information technology. In P. Scrimshaw (Ed.), *Language, classrooms and computers* (pp. 11–26). London: Routledge.

Jones, C. (1999a). Contextualise and personalise: Key strategies for vocabulary acquisition. *ReCALL, 11*(3), 34–40.

Jones, C. (Ed.). (1999b). Language courseware design [Special Issue]. *CALICO Journal, 17*(1).

Jones, C., & Fortescue, S. (1987). *Using computers in the language classroom*. London: Longman Group UK Ltd.

Jones, L. C. (2003). Supporting listening comprehension and vocabulary acquisition with multimedia annotations: The students' voice. *CALICO Journal, 21*(1), 41–65.

Jordan, G. (2004). *Theory construction in second language acquisition.* Amsterdam: John Benjamins.

Kaltenböck, G. (2001). Learner autonomy: A guiding principle in designing a CD-ROM for intonation practice. *ReCALL, 13*(2), 179–190.

Kang, Y.-S., & Maciejewski, A. A. (2000). A student model of technical Japanese reading proficiency for an Intelligent tutoring system. *CALICO Journal, 18*(1), 9–40.

Kawai, G., & Hirose, K. (2000). Teaching the pronunciation of Japanese double-mora phonemes using speech recognition technology. *Speech Communication, 30*(2–3), 131–143.

Kempen, G. (1999). Visual grammar: Multimedia for grammar and spelling instruction in primary education. In K. Cameron (Ed.), *Computer assisted language learning (CALL): Media, design & applications* (pp. 223–238). Lisse, The Netherlands: Swets & Zeitlinger.

Kempfert, T. (2002). Editor's emailbag: Two responses to last month's article on Internet plagiarism. *Teaching with Technology Today, 8*(5). Retrieved July 8, 2005 from http://www.uwsa.edu/ttt/articles/letters.htm#ken

Kiernan, P., & Aizawa, K. (2004). Cell phones in task based learning. Are cell phones useful language learning tools? *ReCALL Journal, 16*(1), 71–84.

Kitade, K. (2000). L2 learners' discourse and SLA theories in CMC: Collaborative interaction in Internet chat. *Computer Assisted Language Learning, 13*(1), 143–166.

Klapper, J. (2003). Taking communication to task? A critical review of recent trends in language teaching. *Language Learning Journal, Summer, 27,* 33–42.

Klassen, J., & Milton, P. (1999). Enhancing English language skills using multimedia: Tried and tested. *Computer Assisted Language Learning, 12*(4), 281–294.

Kluge, D. (2002). Tomorrow's CALL: The future in our hands. In P. Lewis (Ed.), *The changing face of CALL: A Japanese perspective* (pp. 245–268). Lisse, The Netherlands: Swets & Zeitlinger.

Knowles, S. (1992). Evaluation of CALL software: A checklist of criteria for evaluation. *On-CALL, 6*(2), 9–20.

Kohn, K. (1994). Distributive language learning in a computer-based multilingual communication environment. In H. Jung & R. Vanderplank (Ed.), *Proceedings of the 1993 CETall Symposium on the Occasion of the 10th AILA World Congress in Amsterdam* (pp. 31–43). Frankfurt: Peter Lang.

Kohn, K. (2001). Developing multimedia CALL: The *Telos Language Partner* approach. *Computer Assisted Language Learning, 14*(3–4), 251–267.

Kol, S., & Schcolnik, M. (2000). Enhancing screen reading strategies. *CALICO Journal, 18*(1), 67–80.

Kollock, P., & Smith, M. (1996). Managing the virtual commons: Cooperation and conflict in computer communities. In S. Herring (Ed.), *Computer mediated communication: Linguistic, social and cross-cultural perspectives* (pp. 109–128). Amsterdam: John Benjamins.

Komori, S., & Zimmerman, E. (2001). A critique of web-based Kanji learning programs for autonomous learners: Suggestions for improvement of WWKanji. *Computer Assisted Language Learning, 14*(1), 43–67.

Kötter, M. (2003). Negotiation of meaning and codeswitching in online tandems. *Language Learning & Technology, 7*(2), 145–172.

Kötter, M., Shield, L., & Stevens, A. (1999). Real-time audio and email for fluency: Promoting distance language learners' aural and oral skills via the Internet. *ReCALL, 11*(2), 55–60.

Kramsch, C., & Anderson, R. W. (1999). Teaching text and context through multimedia. *Language Learning & Technology, 2*(2), 13–42.

Krashen, S. (1977). The monitor model of adult second language performance. In M. Burt, H. Dulay, & M. Finocchiaro (Eds.), *Viewpoints on English as a second language* (pp. 155–161). New York: Regents.

Krashen, S. (1985) *The input hypothesis: Issues and implications.* Harlow, UK: Longman.

Krathwohl, D. R. (1993). *Methods of educational and social science research: An integrated approach.* New York: Longman.

Kress, G. (2000). Multimodality. In B. Cope & M. Kalantzis (Eds.), *Multiliteracies: Literacy learning and the design of social futures* (pp. 182–202). London: Routledge.

Kress, G., & van Leeuwen, T. (2001). *Multimodal discourse: The modes and media of contemporary communication.* London: Arnold.

Kukulska-Hulme, A. & Traxler, J. (Eds.). (2005). *Mobile technologies for teaching and learning.* London: Kogan Page/Taylor & Francis.

Labrie, G. (2000). A French vocabulary tutor for the web. *CALICO Journal, 17*(3), 475–499.

Lamy, M.-N., & Goodfellow, R. (1999a). "Reflective conversation" in the virtual language classroom. *Language Learning & Technology, 2*(2), 43–61.

Lamy, M.-N., & Goodfellow, R. (1999b). Supporting language students' interactions in web-based conferencing. *Computer Assisted Language Learning, 12*(5), 457–477.

Lantolf, J. P. (1994). Introduction to the special issue. *Modern Language Journal, 78,* 418–420.

Lantolf, J. P. (1996). Review article. SLA theory building: "Letting all the flowers bloom!" *Language Learning, 46*(4), 713–749.

Lantolf, J. P., & Pavlenko, A. (2001). Second language activity theory: Understanding second language learners as people. In P. Breen (Ed.), *Learner contributions to language learning: New directions in research* (pp. 141–158). Harlow, UK: Longman.

Larsen-Freeman, D., & Long, M. H. (1991). *An introduction to second language acquisition research.* New York: Longman.

Laufer, B., & Hill, M. (2000). What lexical information do L2 learners select in a CALL dictionary and how does it affect word retention? *Language Learning & Technology, 4*(2), 58–76.

Laurillard, D., & Marullo, G. (1993). Computer-based approaches to second language learning. In P. Scrimshaw (Ed.), *Language, classrooms and computers* (pp. 145–165). London: Routledge.

Leahy, C. (2001). Bilingual negotiation via e-mail: An international project. *Computer Assisted Language Learning, 14*(1), 15–42.

Lee, L. (1997). Using Internet tools as an enhancement of C2 teaching and learning. *Foreign Language Annals, 30*(3), 410–427.

Lee, L. (2001). Online interaction: Negotiation of meaning and strategies used among learners of Spanish. *ReCALL, 13*(2), 232–244.

Lee, L. (2002a). Enhancing learners' communication skills to synchronous electronic interaction and task based instruction. *Foreign Language Annals, 35*(1), 16–24.

Lee, L. (2002b). Synchronous online exchanges: A study of modification devices on non-native discourse. *System, 30,* 275–288.

Lee, L. (2004). Learners' perspectives on networked collaborative interaction with native speakers of Spanish in the U.S. *Language Learning & Technology, 8*(1), 83–100.

Lehtonen, T., & Tuomainen, S. (2003). CSCL—A tool to motivate foreign language learners: The Finnish application. *ReCALL, 15*(1), 51–67.

Leont'ev, A. N. (1978). *Activity, consciousness and personality.* Englewood Cliffs, NJ: Prentice-Hall.

Leont'ev, A. N. (1981). *Psychology and the language learning process.* Oxford, UK: Pergamon.

Levine, A., Ferenz, O., & Reves, T. (1999). A computer mediated curriculum in the EFL academic writing class. *ReCALL, 11*(1), 72–79.

Levis, J., & Pickering, L. (2004). Teaching intonation in discourse using speech visualization technology. *System, 32*(4), 505–524.

Levy, M. (1993). Integrating CALL into a communicative writing course. *On-CALL, 6*(1), 11–18.

Levy, M. (1997). *Computer-assisted language learning: Context and conceptualisation.* Oxford, UK: Clarendon.

Levy, M. (1999a). Design processes in CALL: Integrating theory, research and evaluation. In K. Cameron (Ed.), *Computer assisted language learning (CALL): Media, design and applications* (pp. 83–108). Lisse, The Netherlands: Swets & Zeitlinger.

Levy, M. (1999b). Theory and design in a multimedia CALL project in cross-cultural pragmatics. *Computer Assisted Language Learning, 12*(1), 29–58.

Levy, M. (2000). Scope, goals and methods in CALL research: Questions of coherence and autonomy. *ReCALL, 12*(2), 170–195.

Levy, M. (2001). Coherence and direction in CALL research. In K. Cameron (Ed.), *CALL—the challenge of change: Research and practice* (pp. 5–14). Exeter, UK: Elm Bank Publications.

Levy, M. (2002). CALL by design: Discourse, products and processes. *ReCALL, 14*(1), 58–84.

Levy, M. (2006). Effective use of CALL technologies: Finding the right balance. In R. Donaldson & M. Haggstrom (Eds.), *Changing language education through CALL* (pp. 1–18). Lisse, The Netherlands: Swets & Zeitlinger.

Levy, M., & Farrugia, D. (1988). *Computers in language teaching: Analysis, research and reviews.* Melbourne, Australia: Footscray College of Technical and Further Education.

Levy, M., & Hubbard, P. (2006). Why call CALL "CALL"? *Computer-Assisted Language Learning, 18*(3), 143–149.

Levy, M., & Kennedy, C. (2004). A task-cycling pedagogy using audio-conferencing and stimulated reflection for foreign language learning. *Language Learning & Technology, 8*(2), 50–68.

Levy, M., & Kennedy, C. (2005). Learning Italian via Mobile SMS. In A. Kukulska-Hulme & J. Traxler (Eds.), *Mobile technologies for teaching and learning* (pp. 76–83). London: Kogan Page/Taylor & Francis.

Li, Y. (2000). Linguistic characteristics of ESL writing in task-based e-mail activities. *System, 28,* 229–245.

Liddicoat, A. J. (2000). Everyday speech as culture: Implications for language teaching. In A. J. Liddicoat & C. Crozet (Eds.), *Teaching languages, teaching cultures* (pp. 51–63). Melbourne, Australia: Applied Linguistics Association of Australia.

Light, P. (1993). Collaborative learning with computers. In P. Scrimshaw (Ed.), *Language, classrooms and computers* (pp. 40–56). London: Routledge.

Ligorio, M. B. (2001). Integrating communication formats: Synchronous versus asynchronous and text-based versus visual. *Computers and Education, 37,* 103–125.

Ligorio, M. B., Talamo, A., & Cesareni, D. (1999, August). *Building the future from the past.* Paper presented at the European Association for Research on Learning and Instruction (EARLI) Conference, Göteborg, Sweden.

Lim, C. P., & Hang, D. (2003). An activity theory approach to research of ICT integration in Singapore schools. *Computers and Education, 41,* 49–63.

Liou, H.-C. (1994). Practical considerations for multimedia courseware development: An EFL IVD experience. *CALICO Journal, 11*(3), 47–74.

Little, D. (2001). Learner autonomy, self-instruction and new technologies in language learning: Current theory and practice in higher education in Europe. In A. Chambers & G. Davies (Eds.), *ICT and language learning: A European perspective* (pp. 29–38). Lisse, The Netherlands: Swets & Zeitlinger.

Little, D., & Ushioda E. (1998). Designing, implementing and evaluating a project in tandem language learning via email. *ReCALL, 10*(1), 95–101.

Long, M. H. (1983). Linguistic and conversational adjustments to non-native speakers. *Studies in Second Language Acquisition, 5*, 177–193.

Long, M. H. (1993). Assessment strategies for second language acquisition theories. *Applied Linguistics, 14*(3), 225–249.

Long, M. H. (1996). The role of the linguistic environment in second language acquisition. In W. C. Ritchie & T. K. Bhatia (Eds.), *Handbook of second language acquisition* (pp. 413–468). San Diego: Academic Press.

Long, M. H., & Crookes, G. (1991). Three approaches to task-based syllabus design. *TESOL Quarterly, 26*, 27–55.

Long, M. H., & Robinson, P. (1998). Focus on form: Theory, research and practice. In C. J. Doughty & J. Williams (Eds.), *Focus on form in classroom second language acquisition* (pp. 15–41). Cambridge, UK: Cambridge University Press.

Loucky, J. P. (2002). Assessing the potential of computerised bilingual dictionaries for enhancing English vocabulary learning. In P. Lewis (Ed.), *The changing face of CALL: A Japanese perspective* (pp. 123–138). Lisse, The Netherlands: Swets & Zeitlinger.

Ma, Q. (2004, September). *Theoretical and design issues for a computer assisted vocabulary learning program: WUFUN.* Paper presented at the Eleventh International CALL Conference, University of Antwerp, Belgium.

Marvin, L.-E. (1995). Spoof, spam, lurk and lag: The aesthetics of text-based virtual realities. *Journal of Computer-Mediated Communication, 1*(2). Retrieved April 22, 2004 from http://www.ascusc.org/jcmc/vol1/issue2/marvin.html

Matthews, C. (1993). Grammar frameworks in Intelligent CALL. *CALICO Journal, 11*(1), 5–27.

Mayer, R. E. (1997). Multimedia learning: Are we asking the right questions? *Educational Psychologist, 32*(1), 1–19.

Mayer, R. E. (2001). *Multimedia learning.* Cambridge, UK: Cambridge University Press.

McDonell, W. (1992). Language and cognitive development through cooperative group work. In C. Kessler (Ed.), *Cooperative language learning* (pp. 51–64). London: Prentice Hall.

Meskill, C. (1999). Computers as tools for sociocollaborative language learning. In K. Cameron (Ed.), *Computer assisted language learning (CALL): Media, design and applications* (pp. 141–164). Lisse, The Netherlands: Swets & Zeitlinger.

Mills, D. (1999). Interactive web-based language learning: The state of the art. In R. Debski & M. Levy (Eds.), *WORLDCALL: Global perspectives on computer-assisted language learning* (pp. 117–132). Lisse, The Netherlands: Swets & Zeitlinger.

Mills, J. (1999). CA-EAP: A multitask software package for the teaching of academic writing. In K. Cameron (Ed.), *CALL & the learning community* (pp. 345–354). Exeter, UK: Elm Bank.

Mishan, F., & Strunz, B. (2003). An application of XML to the creation of an interactive resource for authentic language learning tasks. *ReCALL, 15*(2), 237–250.

Mitchell, R., & Myles, F. (1998). *Second language learning theories.* London: Arnold.

Mitchell, R., & Myles, F. (2004). *Second language learning theories* (2nd ed.). London: Arnold.

Molebash, P. E., Dodge, B., Bell, R. L., Mason, C. L., & Irving, K. E. (2003). Promoting student enquiry: Webquests to web inquiry projects (WIPs). Retrieved July 23, 2004, from http://edweb.sdsu.edu/wip/WIP_intro.htm

Möllering, M. (2000). Computer-mediated communication: Learning German online in Australia. *ReCALL, 12*(1), 27–34.

Moran, C. (1991). We write, but do we read? *Computers and Composition, 8*(3), 51–61.

Mostow, J., & Aist, G. (2001). Evaluating tutors that listen: An overview of project LISTEN. In K. D. Forbus & P. J. Felkovich (Eds.). *Smart machines in education* (pp. 169–234). Menlo Park, CA: AAAI Press.

Moundridou, M., & Virvou, M. (2003). Analysis and design of a web-based authoring tool generating intelligent tutoring systems. *Computers and Education, 40*(2), 157–181.

Mugane, J. (1999). Digital arenas in the delivery of African languages for the development of thought. In R. Debski & M. Levy (Eds.), *WORLDCALL: Global perspectives on computer-assisted language learning* (pp. 33–48). Lisse, The Netherlands: Swets & Zeitlinger.

Murray, D. E. (2000). Protean communication: The language of computer-mediated communication. *TESOL Quarterly, 34*(3), 397–421.

Murray, G. L. (1999a). Autonomy and language learning in a simulated environment. *System, 27*(3), 295–308.

Murray, G. L. (1999b). Exploring learners' CALL experiences: A reflection on method. *Computer Assisted Language Learning, 12*(3), 179–195.

Murray, L., & Barnes, A. (1998). Beyond the "wow" factor—Evaluating multimedia language learning software from a pedagogical viewpoint. *System, 26*, 249–259.

Nagata, N. (2002). BANZAI: An application of natural language processing to web-based language learning. *CALICO Journal, 19*(3), 583–600.

Nation, I. S. P. (2001). *Learning vocabulary in another language.* Cambridge, UK: Cambridge University Press.

Negretti, R. (1999). Web-based activities and SLA: A conversation analysis research approach. *Language Learning & Technology, 3*(1), 75–87.

Nelson, T., & Oliver, W. (1999). Murder on the Internet. *CALICO Journal, 17*(1), 101–114.

Neri, A., Cucchiarini, C., Strik, H., & Boves, L. (2002). The pedagogy–technology interface in computer-assisted pronunciation training. *Computer Assisted Language Learning, 15*(5), 441–468.

Nesselhauf, N., & Tschichold, C. (2002). Collocations in CALL: An investigation of vocabulary building software for EFL. *Computer Assisted Language Learning, 15*(3), 251–280.

Neuman, W. L. (2003). *Social research methods: Qualitative and quantitative approaches* (5th ed.). Boston: Allyn & Bacon.

Neumeyer, L., Franco, H., Digalakis, V., & Weintraub, M. (2000). Automatic scoring of pronunciation quality. *Speech Communication, 30*(2–3), 83–93.

Ng, K. L. E., & Olivier, W. (1987). Computer-assisted language learning: An investigation on some design and implementation issues. *System, 15*(1), 1–17.

Nikolova, O. (2002). Effects of students' participation in authoring of multimedia materials on student acquisition of vocabulary. *Language Learning & Technology, 6*(1), 100–122.

Norris, J., & Ortega, L. (2003). Defining and measuring SLA. In C. J. Doughty & M. H. Long (Eds.), *The handbook of second language acquisition* (pp. 717–761). Oxford, UK: Basil Blackwell.

Nunan, D. (1989). *Designing tasks for the communicative classroom.* Cambridge, UK: Cambridge University Press.

Nunan, D. (1992). *Research methods in language learning.* Cambridge, UK: Cambridge University Press.

Nunan, D. (2004). *Task-based language teaching.* Cambridge, UK: Cambridge University Press.

Nunan, D. (Ed.). (2005). Technology and oral language development. [Special Issue]. *Language Learning & Technology, 9*(3).

O'Dowd, R. (2003). Understanding the "other side": Intercultural learning in a Spanish–English e-mail exchange. *Language Learning & Technology, 7*(2), 118–144.

Papert, S. (1993). *The children's machine: Rethinking school in the age of the computer.* New York: Basic Books.

Paramskas, D. M. (1999). The shape of computer-mediated communication. In K. Cameron (Ed.), *Computer assisted language learning (CALL): Media, design and applications* (pp. 13–34). Lisse, The Netherlands: Swets & Zeitlinger.

Parks, S., Huot, D., Hamers, J., & H.-Lemonnier, F. (2003). Crossing boundaries: Multimedia technology and pedagogical innovation in high school class. *Language Learning & Technology, 7*(1), 28–45.

Patrikis, P. C. (1997). The evolution of computer technology in foreign language teaching and learning. In R. Debski, J. Gassin, & M. Smith (Eds.), *Language learning through social computing* (Applied Linguistics of Australia Occasional Papers Number 16; pp. 159–177). Parkville, Australia: Applied Linguistics Association of Australia.

Payne, J. S., & Whitney, P. J. (2002). Developing L2 oral proficiency through synchronous CMC: Output, working memory, and interlanguage development. *CALICO Journal, 20*(1), 7–32.

Pecorari, D. (2003). Good and original: Plagiarism and patchwriting in academic second-language writing. *Journal of Second Language Writing, 12*, 317–345.

Pederson, M. (1988). Research in CALL. In Wm. Flint Smith (Ed.), *Modern media in foreign language education: Theory and implementation* (pp. 99–132). Lincolnwood, IL: National Textbook.

Pelgrum, W. J. (2001). Obstacles to the integration of ICT in education: Results from a worldwide educational assessment. *Computers & Education, 37*, 163–178.

Pellettieri, J. (2000). Negotiation in cyberspace: The role of chatting in the development of grammatical competence. In M. Warschauer & R. Kern (Eds.), *Network-based language teaching: Concepts and practice* (pp. 59–86). Cambridge, UK: Cambridge University Press.

Pennington, M. (1999a). Computer aided pronunciation pedagogy: Promise, limitations, directions. *Computer Assisted Language Learning, 12*(5), 427–440.

Pennington, M. (1999b). The missing link in computer-assisted writing. In K. Cameron (Ed.), *Computer assisted language learning (CALL): Media, design and applications* (pp. 271–294). Lisse, The Netherlands: Swets and Zeitlinger.

Pennington, M. (2004). Electronic media in second language writing: An overview of tools and research findings. In S. Fotos & C. M. Browne (Eds.), *New perspectives on CALL for second language classrooms* (pp. 69–92). Mahwah, NJ: Lawrence Erlbaum Associates.

Pérez, L. C. (2003). Foreign language productivity in synchronous versus asynchronous computer-mediated communication. *CALICO Journal, 21*(1), 89–104.

Peterson, M. (2001). MOOs and second language acquisition: Towards a rationale for MOO-based learning. *Computer Assisted Language Learning, 14*(5), 443–459.

Phillips, D. C. (1995). The good, the bad, and the ugly: The many faces of construct-ivism. *Educational Researcher, 24*(7), 5–12.

Phillips, M. (1985a). Educational technology in the next decade: An ELT perspec-tive. In C. Brumfit, M. Phillips, & P. Skehan, (Eds.), *Computers and English lan-guage teaching: ELT Documents 122* (pp. 99–119). Oxford, UK: Pergamon.

Phillips, M. (1985b). Logical possibilities and classroom scenarios for the develop-ment of CALL. In C. Brumfit, M. Phillips, & P. Skehan (Eds.), *Computers and Eng-lish language teaching: ELT Documents 122* (pp. 120–159). Oxford, UK: Pergamon.

Pica, T. (1991). Classroom interaction, participation and comprehension: Redefin-ing relationships. *System, 19*(3), 437–452.

Postman, N. (1993). *Technopoly: The surrender of culture to technology.* New York: Vintage.

Precoda, K. (2004). Non-mainstream languages and speech recognition: Some chal-lenges. *CALICO Journal, 21*(2), 229–243.

Preece, J., Rogers, Y., Sharp, H., Benyon, D., Holland, S., & Carey, T. (1994). *Human-computer interaction.* Workingham, UK: Addison-Wesley Publishing Company.

Price, C., McCalla, G., & Bunt, A. (1999). L2 tutor: A mixed-initiative dialogue sys-tem for improving fluency. *Computer Assisted Language Learning, 12*(2), 83–112.

Price, S., & Rogers, Y. (2004). Let's get physical: The learning benefits of interacting in digitally augmented physical spaces. *Computers and Education, 43*, 137–151.

Pujolà, J.-T. (2001). Did CALL feedback feed back? Researching learners' use of feedback. *ReCALL, 13*(1), 79–98.

Pujolà, J.-T. (2002). CALLing for help: Research in language learning strategies us-ing help facilities in a web-based multimedia program. *ReCALL, 14*(2), 235–262.

Pusack, J. P. (1983). Answer processing and error correction in foreign language CAI. *System, 11*(1), 53–64.

Pusack, J. P. (1999). The Kontakte multimedia project at the University of Iowa. *CALICO Journal, 17*(1), 25–42.

Ribé, R., & Vidal, N. (1993). *Project work: Step by step.* Oxford, UK: Heinemann.

Richards, J. C., & Rodgers, T. S. (1986). *Approaches and methods in language teaching.* Cambridge, UK: Cambridge University Press.

Robinson, G. (1991). Effective feedback strategies in CALL: Learning theory and empirical research. In P. Dunkel (Ed.), *Computer-assisted language learning and test-ing* (pp. 155–167). New York: Newbury House.

Robinson, P. (Ed.). (2002). *Individual differences and instructed language learning.* Phila-delphia/Amsterdam: John Benjamins.

Rodgers, E. M. (2003). *Diffusion of innovations* (5th ed.). New York: Free Press.

Roed, J. (2003). Language learner behaviour in a virtual environment. *Computer As-sisted Language Learning, 16*(2–3), 155–173.

Rogerson-Revell, P. (2003). Developing a cultural syllabus for business language e-learning materials. *ReCALL, 15*(2), 155–168.

Roschelle, J. (2003). Keynote paper: Unlocking the learning value of wireless mobile devices. *Journal of Computer Assisted Learning, 19*, 260–272.

Rosell-Aguilar, F. (2004). WELL done and well liked: Online information literacy skills and learner impressions of the web as a resource for foreign language learn-ing. *ReCALL, 16*(1), 210–224.

Rüschoff, B., & Ritter, M. (2001). Technology enhanced language learning: Con-struction of knowledge and template based learning in the foreign language classroom. *Computer Assisted Language Learning, 14*(3–4), 219–232.

Ryan, J. (1998). *Student plagiarism in an online world.* ASEE PRISM Online. Retrieved January 26, 2005 from http://www.asww.org/prism/december/html/student_plagiarism_in_an_online.htm

Saarenkunnas, M., Kuure, L., & Taalas, P. (2003). The poly contextual nature of computer supported learning—Theoretical and methodological perspectives. *Re-CALL, 15*(2), 202–216.

Saita, I., Harrison, R., & Inman, D. (1998) Learner proficiency and learning tasks using email in Japanese. In K. Cameron (Ed.), *Multimedia CALL: Theory and practice* (pp. 221–228). Exeter, UK: Elm Bank.

Salaberry, M. R. (1996). The theoretical foundation for the development of pedagogical tasks in computer mediated communication. *CALICO Journal, 14*(1), 5–34.

Salaberry, M. R. (2000a). L2 morphosyntactic development in text-based computer-mediated communication. *Computer Assisted Language Learning, 13*(1), 5–27.

Salaberry, M. R. (2000b). Pedagogical design of computer-mediated communication tasks: Learning objectives and technological capabilities. *The Modern Language Journal, 84*(1), 28–37.

Salomon, G. (1991). Transcending the qualitative–quantitative debate: The analytic and systemic approaches to educational research. *The Educational Researcher, 20*(6), 10–18.

Schmidt, R. (1990). The role of consciousness in second language learning. *Applied Linguistics, 11*(2), 129–158.

Schmidt, R. (1994). Deconstructing consciousness in search of useful definitions for applied linguistics. *AILA Review, 11*, 11–26.

Schofield, J. (1995). *Computers and classroom culture.* Cambridge, UK: Cambridge University Press.

Schultz, M. (2000). Computers and collaborative writing in the foreign language curriculum. In M. Warschauer & R. Kern (Eds.), *Network-based language teaching: Concepts and practice* (pp. 121–150). Cambridge, UK: Cambridge University Press.

Schulze, M. (2001). Human language technologies in Computer-assisted language learning. In A. Chambers & G. Davies (Eds.), *ICT and language learning: A European perspective* (pp. 111–132). Lisse, The Netherlands: Swets & Zeitlinger.

Schulze, M. (2003). Grammatical errors and feedback: Some theoretical insights. *CALICO Journal, 20*(3), 437–450.

Schwienhorst, K. (2002). Evaluating tandem language learning in the MOO: Discourse repair strategies in a bilingual Internet project. *Computer Assisted Language Learning, 15*(2), 135–145.

Schwienhorst, K., & Kapec, P. (2003, September). *Balancing bilingualism in MOO tandem: Learner attitudes to the bilingual tandem analyser.* Paper presented at the Eurocall 2003 Conference, University of Limerick, Ireland.

Sengupta, S. (2001). Exchanging ideas with peers in network-based classrooms: An aid or a pain? *Language Learning & Technology, 5*(1), 103–134.

Sharwood-Smith, M. (1993). Input enhancement in instructed SLA: Theoretical bases. *Studies in Second Language Acquisition, 15*(2), 165–180.

Shaughnessy, M. (2003). CALL, commercialism, and culture: Inherent software design conflicts and their results. *ReCALL, 15*(2), 251–268.

Shawback, M. J., & Terhune, N. M. (2002). Online interactive courseware: Using movies to promote cultural understanding in a CALL environment. *ReCALL, 14*(1), 85–95.

Shield, L. (2003). MOO as a language learning tool. In U. Felix (Ed.), *Language learning online: Towards best practice* (pp. 97–122). Lisse, The Netherlands: Swets & Zeitlinger.

Shield, L., Weininger, M., & Davies, L. B. (1999). A task-based approach to using MOO for collaborative language learning. In K. Cameron (Ed.), *CALL & the learning community* (pp. 391–402). Exeter, UK: Elm Bank.

Shimatani, H., & Stockwell, G. R. (2003). An evaluation of a self-study CALL environment: Language development learner attitudes. *Language Education and Technology, 40,* 1–14.

Shin, J., & Wastell, D. G. (2001). A user-centred methodological framework for the design of hypermedia based CALL systems. *CALICO Journal, 18*(3), 517–537.

Shneiderman, B. (1987). *Designing the user interface: Strategies for effective human–computer interaction.* Reading, MA: Addison-Wesley.

Sivert, S., & Egbert, J. (1999). CALL issues: Building a computer-enhanced language classroom. In J. Egbert & E. Hanson-Smith (Eds.), *CALL environments: Research, practice and critical issues* (pp. 41–49). Alexandria, VA: TESOL.

Skehan, P. (1998). *The cognitive approach to language learning.* Oxford, UK: Oxford University Press.

Skehan, P. (2003). Focus on form, tasks and technology. *Computer Assisted Language Learning, 16*(5), 391–411.

Skehan, P., & Foster, P. (1997). The influence of planning and post-task activities on accuracy and complexity in task-based learning. *Language Teaching Research, 1*(3), 185–211.

Skehan, P., & Foster, P. (2001). Cognition and tasks. In P. Robinson (Ed.), *Cognition and second language instruction* (pp. 183–205). Cambridge, UK: Cambridge University Press.

Smith, B. (2003). Computer-mediated negotiated interaction: An expanded model. *The Modern Language Journal, 87*(1), 38–57.

Smith, B., &. Gorsuch, G. J. (2004). Synchronous computer mediated communication captured by usability lab technologies: New interpretations. *System, 32*(4), 553–575.

Soboleva, O., & Tronenko, N. (2002). A Russian multimedia learning package for classroom use and self-study. *Computer Assisted Language Learning, 15*(5), 483–500.

Söntgens, K. (1999). Language learning via email—autonomy through collaboration. In K. Cameron (Ed.), *CALL & the learning community* (pp. 413–424). Exeter, UK: Elm Bank.

Söntgens, K. (2001). Circling the globe: Fostering experiential language learning. *ReCALL, 13*(1), 59–66.

Sotillo, S. M. (2000). Discourse functions and syntactic complexity in synchronous and asynchronous communication. *Language Learning & Technology 4*(1), 82–119.

Spada, N. (1997). Form-focused instruction and second language acquisition. *Language Teaching, 30*(2), 73–87.

Stauffer, S. J. (1994). Computer-based classrooms for language teaching. In *Georgetown University roundtable on languages and linguistics* (pp. 220–232). Washing, DC: Georgetown University Press.

Stepp-Greany, J. (2002). Student perceptions on language learning in a technological environment: Implications for the new millennium. *Language Learning & Technology, 6*(1), 165–180.

Stevens, A., & Hewer, S. (1998). From policy to practice and back. In the proceedings of the 1st LEVERAGE Conference, Cambridge. 7-8 January, 1998. Retrieved on June 18, 2004 from http://greco.dit.upm.es/~leverage/conf1/hewer.htm

Stevens, V. (2002). Concordance, version 2.0. *CALICO Journal, 19*(3), 691–705.

St. John, E. (2001). A case for using parallel corpus and concordancer for beginners of a foreign language. *Language Learning & Technology, 5*(3), 185–203.

Stockwell, G. R. (2003a). Effects of topic threads on sustainability of email interactions between native speakers and nonnative speakers. *ReCALL, 15*(1), 37.

Stockwell, G. R. (2003b). What do learners acquire through email interactions with native speakers? *LET Kyushu–Okinawa Bulletin, 3,* 31–42.

Stockwell, G. R. (2004). Communication breakdown in asynchronous CMC. *Australian Language and Literacy Matters, 1*(3), 7–10, 31.

Stockwell, G. R., & Harrington, M. W. (2003). The incidental development of L2 proficiency in NS-NNS email interactions. *CALICO Journal, 20*(2), 337–359.

Stockwell, G. R., & Levy, M. (2001). Sustainability of email interactions between native speakers and nonnative speakers. *Computer Assisted Language Learning, 14*(5), 419–442.

Stockwell, G. R., & Nozawa, K. (2004, September). *Total integration? Using CALL in an ESP IT curriculum.* Paper presented at the EuroCALL 2004 Conference, University of Vienna, Austria.

Strambi, A., & Bouvet, E. (2003). Flexibility and interaction at a distance: A mixed-mode environment for language learning. *Language Learning & Technology, 7*(3), 81–102.

Susser, B. (2001). The defence of checklists for courseware evaluation. *ReCALL, 13*(2), 261–276.

Svensson, P. (2003). Virtual worlds as arenas for language learning. In U. Felix (Ed.), *Language learning online: Towards best practice* (pp. 123–142). Lisse, The Netherlands: Swets & Zeitlinger.

Swain, M. (1985). Communicative competence: Some roles of comprehensible input and comprehensible output in its development. In S. Gass & C. Madden (Eds.), *Input in second language acquisition* (pp. 235–253). Rowley, MA: Newbury House.

Swain, M., & Lapkin, S. (1995). Problems in output and the cognitive processes they generate: A step toward. *Applied Linguistics, 16*(3), 371–391.

Taylor, R. P., & Gitsaki, C. (2003). Teaching WELL in a computerless classroom. *Computer Assisted Language Learning, 16*(4), 275–294.

Taylor, R. P., & Gitsaki, C. (2004). Teaching WELL and loving IT. In S. Fotos & C. M. Browne (Eds.), *New perspectives on CALL for second language classrooms* (pp. 131–148). Mahwah, NJ: Lawrence Erlbaum Associates.

Thornton, P., & Houser, C. (2001). Learning on the move: Foreign language vocabulary via SMS. *ED-Media 2001 proceedings* (pp. 1846–1847). Norfolk, VA: Association for the Advancement of Computing in Education.

Thornton, P., & Houser, C. (2002). M-learning: Learning in transit. In P. Lewis (Ed.), *The changing face of CALL: A Japanese perspective* (pp. 229–243). Lisse, The Netherlands: Swets & Zeitlinger.

Tokuda, N., & Chen, L. (2004). A new KE-free online ICALL system featuring error contingent feedback. *Computer Assisted Language Learning, 17*(2), 177–201.

Tolmie, A., & Boyle, J. (2000). Factors influencing the success of computer mediated communication (CMC) environments in university teaching: A review and case study. *Computers and Education, 34,* 119–140.

Tomlinson, B. (Ed.). (1998). *Materials development in language teaching.* Cambridge, UK: Cambridge University Press.

Toole, J., & Heift, T. (2002). The tutor assistant: An authoring tool from intelligent language tutoring system. *Computer Assisted Language Learning, 15*(4), 373–386.

Toyoda, E., & Harrison, R. (2002). Categorization of text chat communication between learners and native speakers of Japanese. *Language Learning & Technology, 6*(1), 82–99.

Trinder, R. (2003). Conceptualisation and development of multimedia courseware in the tertiary educational context: Juxtaposing approach, content and technology considerations. *ReCALL, 15*(1), 79–93.

Tschichold, C. (1999). Grammar checking for CALL: Strategies for improving foreign language grammar checkers. In K. Cameron (Ed.), *Computer assisted language learning (CALL): Media, design and applications* (pp. 203–222). Lisse, The Netherlands: Swets & Zeitlinger.

Tschirner, E. (2001). Language acquisition in the classroom: The role of digital video. *Computer Assisted Language Learning, 14*(3–4), 305–320.

Tsou, W., Wang, W., & Li, H.-Y. (2002). How computers facilitate English foreign language learners acquire English abstract words. *Computers and Education, 39*(4), 415–428.

Tsubota, Y., Dantsuji, M., & Kawahara, T. (2004). An English pronunciation learning system for Japanese students based on diagnosis of critical pronunciation errors. *ReCALL, 16*(1),173–188.

Tudini, V. (2003). Using native speakers in chat. *Language Learning & Technology, 7*(3), 141–159.

Tudini, V. (2004). Virtual immersion: Native speaker chats as a bridge to conversational Italian. *Australian Review of Applied Linguistics, Series S, 18,* 63–80.

Turnbull, J. (2002). WordPilot 2000 premium edition (Speech Pilot). *CALICO Journal, 19*(3), 673–687.

Van de Poel, K., & Swanepoel, P. (2003). Theoretical and methodological pluralism in designing effective lexical support for CALL. *Computer Assisted Language Learning, 16*(2–3), 109–134.

Vandeventer, A. (2001). Creating a grammar checker for CALL by constraint relaxation: A feasibility study. *ReCALL, 13*(1), 110–120.

Vannatta, R. A., & Beyerbach, B. (2000). Facilitating the constructivist vision of technology integration among education faculty and preservice teachers. *Journal of Research on Technology in Education, 33*(2), 132–148.

Vanparys, J., & Baten, L. (1999). How to offer real help to grammar learners. *ReCALL, 11*(1), 125–132.

Varonis, E. M., & Gass, S. M. (1985). Non-native/non-native conversations: A model for negotiation of meaning. *Applied Linguistics, 6*(1), 71–90.

Vick, R. M., Crosby, M. E., & Ashworth, D. E. (2000). Japanese and American students meet on the web: Collaborative language learning through everyday dialogue with peers. *Computer Assisted Language Learning, 13*(3), 199–219.

Vila, J., Lim, B., & Anajpure, A. (2004). VOWELS: A voice-based web engine for locating speeches. *Journal of Educational Multimedia and Hypermedia, 13*(2), 129–141.

Vygotsky, L. S. (1978). *Mind in society.* Cambridge, MA: Harvard University Press.

Wachowicz, K., & Scott, B. (1999). Software that listens: It's not a question of whether, it's a question of how. *CALICO Journal, 6*(3), 253–276.

Wang, L. (2001). Exploring parallel concordancing in English and Chinese. *Language Learning & Technology, 5*(3),174–184.

Wang, S., Higgins, M., & Shima, Y. (2005). Training English pronunciation for Japanese learners of English online. *The JALT CALL Journal, 1*(1), 39–47.

Wang, X., & Munro, M. J. (2004). Computer-based training for learning English vowel contrasts. *System, 32*(4), 539–552.

Wang, Y. (2004a). Distance language learning: Interactivity and fourth-generation Internet-based videoconferencing. *CALICO Journal, 21*(2), 373–396.

Wang, Y. (2004b). Supporting synchronous distance language learning with desktop videoconferencing. *Language Learning & Technology, 8*(3), 90–121.

Wang, Y., & Sun, C. (2001). Internet-based real time language education: Towards a fourth generation distance education. *CALICO Journal, 18*(3), 539–561.

Ward, M. (2002). Reusable XML technologies and the development of language learning materials. *ReCALL Journal, 14*(2), 283–292.

Warschauer, M. (Ed.). (1995a). *Computer-mediated collaborative learning: Theory and practice* (NFLRC Research Notes #17). Honolulu, HI: Second Language Teaching & Curriculum Center.

Warschauer, M. (Ed.). (1995b). *Virtual connections: Online activities and projects for networking language learners.* Honolulu, HI: Second Language Teaching & Curriculum Center.

Warschauer, M. (1996). Comparing face-to-face and electronic discussion in the second language classroom. *CALICO Journal, 13*(3), 7–26.

Warschauer, M. (1997). Computer-mediated collaborative learning: Theory and practice. *The Modern Language Journal, 81*(4), 470–481.

Warschauer, M. (2000a). On-line learning in second language classrooms: An ethnographic study. In M. Warschauer & R. Kern (Eds.), *Network-based language teaching: Concepts and practice* (pp. 41–58). Cambridge, UK: Cambridge University Press.

Warschauer, M. (2000b). The death of cyberspace and the rebirth of CALL. *English Teachers' Journal, 53*, 61–67.

Warschauer, M., & Kern, R. (Eds.). (2000). *Network-based language teaching: Concepts and practice.* Cambridge, UK: Cambridge University Press.

Weinberg, A. (2002). Virtual misadventures: Technical problems and student satisfaction when implementing multimedia in an advanced French listening comprehension course. *CALICO Journal, 19*(2), 331–358.

Weinberg, A., & Knoerr, H. (2003). Learning French pronunciation: Audiocassettes or multimedia? *CALICO Journal, 20*(2), 315–336.

Weinholtz, D., Kacer, B., & Rocklin, T. (1995). Pearls, pith and provocation: Salvaging quantitative research with qualitative data. *Qualitative Health Research, 5*(3), 388–397.

Weininger, M., & Shield, L. (2001). Orality in MOO: Rehearsing speech in text: A preliminary study. In K. Cameron (Ed.), *CALL—The challenge of change* (pp. 89–96). Exeter, UK: Elm Bank.

Weininger, M., & Shield, L. (2004, September). *Proximity and distance: A theoretical model for the description and analysis of online discourse.* Paper presented at the Eleventh International CALL Conference, University of Antwerp, Belgium.

Wertsch, J. V. (1998). *Mind as action.* New York: Oxford University Press.

Wible, D., Kuo, C.-H., Chien, F.-Y., Liu, A., & Tsao, N.-L. (2001). A web-based EFL writing environment: Integrating information for learners, teachers and researchers. *Computers and Education, 37*, 297–315.

Wilkins, D. A. (1972). *Linguistics in language teaching.* London: Arnold.

Wilkins, D. A. (1976). *Notional syllabuses.* Oxford, UK: Oxford University Press.

Wilks, Y. (2004). *Machine translation: Its scope and limits.* Cambridge, UK: Cambridge University Press.

Willis, J. (1996). *A framework for task-based learning.* Harlow, UK: Longman.

Winkler, B. (2001). English learners' dictionaries on CD-ROM as reference and language learning tools. *ReCALL, 13*(2), 191–205.

Winograd, T., & Flores, F. (1986). *Understanding computers and cognition.* Reading, MA: Addison-Wesley.

Wood, A. F., & Smith, M. (2005). *Online communication: Linking technology, identity and culture* (2nd ed). Mahwah, NJ: Lawrence Erlbaum Associates.

Wood, J. (2001). Can software support children's vocabulary development? *Language Learning & Technology, 5*(1), 166–201.

Woodin, J. (1997). Email tandem learning and the communicative curriculum. *ReCALL Journal, 9*(1), 22–33.

Wray, R. (2005, July 2). Web challenge to English supremacy. *Guardian Weekly: Learning English,* p. 2.

Xie, T. (2002). Using Internet relay chat in teaching Chinese. *CALICO Journal, 19*(3), 513–524.

Yeh, Y., & Wang, C.-W. (2003). Effects of multimedia vocabulary annotations and learning styles on vocabulary learning. *CALICO Journal, 21*(1), 131–144.

Yoshii, M., & Flaitz, J. (2002). Second language incidental vocabulary retention: The effect of picture and annotation types. *CALICO Journal, 20*(1), 33–58.

Ypsilandis, G. S. (2002). Feedback in distance education. *Computer Assisted Language Learning, 15*(2), 167–182.

Yuan, Y. (2003). The use of chat rooms in an ESL setting. *Computers and Composition, 20*(2), 194–206.

Zhang, H. (2002). Teaching business Chinese online. *CALICO Journal, 19*(3), 525–532.

Author Index

Subject Index

A

Abbreviations, use in chat, 103
Accessibility (for users), 78
Action research, 235
ActionCatcher, 153
Active Worlds, 92
Activity theory, 118–122, 164
    application in context of CMC use, 119–120
    application to HCI, 120
    basis of, 118–119
    concept of shared purpose, 119–120
    implications for language learning, 121
    implications for use of ICT in education, 119
    Leont'ev's analysis of activity, 118
    relationship between CALL and development of learner autonomy, 120–121
Affordances, 97, 191–192
Airline Talk 2, 58
Airline Talk Project, 57
Annotation design, choosing the most effective, experimental research study, 155–157, 173–175, 180, 243
Artifact building, 43
Artificial intelligence (AI), 212–213, 223
Aspects, 90
ASR, *see* automated speech recognition

Assessing student attitudes to web-enhanced learning course, survey research, 146–147, 160, 161
Asynchronous BBS, quantity of language produced, 98
Asynchronous CMC, 84–86, 95
    applicability to language learning, 182, 184, 192
    characteristics, 100–101, 107–108
    quantity and quality of language produced, 98, 99, 102–103
    student effective use of, 193
Audience, for evaluation and research studies, 42
Audience needs, goals and characteristics, designers understanding of 27, 36–37
Audioconferencing, 93, 94, 108, 115, 210–211
    creating opportunities for language learners, 222–223
    difficulties for language learning, 94, 181–182
    for listening, 180–181
    for speaking, 181
Audiographic conferencing tools, 26, 134
    evaluation 54–56
Authenticity (Chapelle criterion), 65
Authenticity-centered approach, 16
Authoring tools, 4, 5, 35, 207–208, 209, 220–221
    for course and syllabus design, 19